Quality Assurance in Analytical Chemistry

Bernd W. Wenclawiak • Michael Koch
Evsevios Hadjicostas
Editors

Quality Assurance in Analytical Chemistry

Training and Teaching

Second Edition

 Springer

Editors
Prof. Dr. Bernd W. Wenclawiak
Universität Siegen
FB 8 Chemie
Inst. Analytische Chemie 1
Adolf-Reichwein-Str. 2
57068 Siegen
Germany
wenclawiak@chemie.uni-siegen.de

Dr. Michael Koch
Universität Stuttgart
Inst. Siedlungswasserbau
Wassergüte-und
Abfallwirtschaft
Abt. Chemie
Bandtäle 1
70569 Stuttgart
Germany
michael.koch@iswa.uni-stuttgart.de

Dr. Evsevios Hadjicostas
Quintessence Enterprises Ltd.
Kennedy Business Center
Office 208
12-14 Kennedy Avenue
1087 Nicosia
Cyprus
quintessence@cytanet.com.cy

The terms and definition taken from ISO 9004:2000, Fig. 1, Quality management systems-guidelines for performance improvements, are reproduced with the permission of the International Organization for Standardization, ISO. This standard can be obtained from any ISO member and from the Web site of the ISO Cental Secretariat with the following address: www.iso.org. Copyright remains with ISO.

Additional material to this book can be downloaded from http://extras.springer.com

ISBN 978-3-642-44851-5   ISBN 978-3-642-13609-2 (eBook)
DOI 10.1007/978-3-642-13609-2
Springer Heidelberg Dordrecht London New York

Cover design: WMXDesign GmbH, Heidelberg, Germany

Printed on acid-free paper

Springer is part of Springer Science+Business Media (www.springer.com)

Foreword to the Second Edition

The first edition of this book came out 2004 and it has been proven very popular with over 1,000 copies sold. With the rapid changes in this field and the publication of the new standard in terminology ISO Guide 99 (VIM3) a decision was taken to make an update. All chapters have been revised in order to follow the terminology in VIM3. The main work of the update was performed by Michael Koch. In this edition also two contributors have taken part in the work, Michael Gluschke and Bertil Magnusson. The number of slides has increased from 756 to slightly more than 800 and the slides in the accompanied electronic material are now available in both English and German. A programme for control charts was added to the electronic material.

Important chapter updates:

- Measurement uncertainty: Since 2004 there has been considerable development in approaches to estimation of uncertainty and this chapter has been considerable revised and expanded in order to take into account new guidelines. Main difference is that several ways of estimating measurement uncertainty are know full acceptable and the analyst is free to choose approach dependent on scope and data availability.
- Calibration: Considerable feedback showed that there was room for improvement. The chapter has been fully revised based on this feedback from readers which we here would like to acknowledge.
- Validation of analytical methods: We all know that validation is to assess fitness for intended purpose. It was therefore logical to combine the separate chapters on Fit for purpose and Validation in the first edition into one chapter.

Bertil Magnusson

Foreword to the First Edition

The application of Quality Assurance (QA) techniques has led to major improvements in the quality of many products and services. Fortunately these techniques have been well documented in the form of guides and standards and nowhere more so than in the area of measurement and testing, particularly chemical analysis.

Training of analysts and potential analysts in quality assurance techniques is a major task for universities and industrial and government laboratories. Re-training is also necessary since the quest for improvements in quality seems to be never ending.

The purpose of this book is to provide training material in the convenient form of PowerPoint slides with notes giving further details on the contents of the slides. Experts in the relevant topic, who have direct experience of lecturing on or utilising its contents, have written each chapter. Almost every aspect of QA is covered from basic fundamentals such as statistics, uncertainty and traceability, which are applicable to all types of measurement, through specific guidance on method validation, use of reference materials and control charts. These are all set in the context of total quality management, certification and accreditation. Each chapter is intended to be self-contained and inevitably this leads to some duplication and cross-references are given if there is more detailed treatment in other chapters.

The accompanying CD contains over 700 PowerPoint slides, which can be used for presentations without any or with little modification and there are extensive lists of references to the guides and standards that can be used to amplify the notes given with each slide. The use of the material in this book should considerably reduce the time and effort needed to prepare presentations and training material.

Alex Williams

Preface and Introduction

The importance of quality assurance of chemical measurements not only for global trade but also for a global society has been characterized in a statement by *Paul de Bièvre*, one of the forerunners concerned about analytical results and their use in widespread applications:

> Chemical measurements are playing a rapidly expanding role in modern society and increasingly form the basis of important decisions.

Acceptability of food is dependent on a knowledge of its ingredients e.g. how pure is the drinking water or is there acrylamide in french fries or other fried food preparations, how much vitamin C, or ß-carotene, or proline is there in juices, what preservatives are there in bread, sausages or other food preparations? Alloys have to meet certain specifications to be used in tools, machinery or instruments. The price of platinum ores or used catalytic converters from cars depends on the platinum content. There are many more examples. This shows the importance of correct analytical results.

The question is: Why are correct analytical results so important today?

The following statements help to understand why:

For correct decisions one needs regulations (e.g. ISO standards).

- Regulations mean limits have to be set and controlled.
- Regulations have an impact on commercial, legal or environmental decisions.
- Quality of traded goods depends on measurements that in turn can be trusted. (Measurements have to be of good quality and reliable.)
- Good measurements require controllable and internationally accepted and agreed procedures.

High quality measurements require qualified specialists. A specialist needs not necessarily a university degree in chemistry. Anyone who is well trained and familiar with the field can become a specialist. However specialists need re-training and their knowledge updating on a regular basis. To help with understanding the different topics involved and to provide a sound basis for quality assurance in an analytical laboratory and also to provide material for teaching and (self) training we have compiled a series of chapters by different authors covering the most important topics. The transparencies are intended for teaching purposes but might also be suitable to give an overview of the subject. We hope that our work will reduce the burden of finding all this information yourselves. All information in this edition has been updated or corrected to the best of our knowledge. This material provided has been collected from different sources. One important source is the material available from EURACHEM.

Eurachem is a network of organisations in Europe having the objective of establishing a system for the international trace-ability of chemical measurements and the promotion of good quality practices. It provides a forum for the discussion of common problems and for developing an informed and considered approach to both technical and policy issues. It provides a focus for analytical chemistry and quality related issues in Europe.

You can find more information about EURACHEM on the internet via "Eurachem –A Focus for Analytical Chemistry in Europe" (http://www.eurachem.org). In particular the site Guides and Documents contains a number of different guides, which might help you to set up a quality system in your laboratory.

The importance of quality assurance in analytical chemistry can best be described by the triangles depicted in Figs. 1 and 2. Quality is checked by testing and testing guaranties good quality. Both contribute to progress in QA (product control and quality) and thus to establishing a market share. Market success depends on quality, price, and flexibility. All three of them are interconnected.

Before you can analyse anything the sample must be taken by someone. This must be of major concern to any analytical chemist. There is no accurate analysis without proper sampling. For correct sampling you need a clear problem definition.

There is no correct sampling without a clear problem definition

Because the sampling error is usually the biggest error in the whole analysis, care must be taken to consider all aspects from sampling. Measurement uncertainties arising from the process of sampling and the physical preparation of the sample can be estimated.

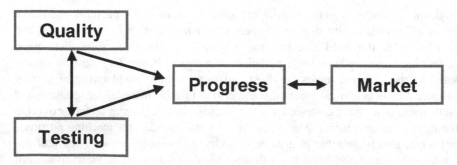

Fig. 1 Factors that influence the market

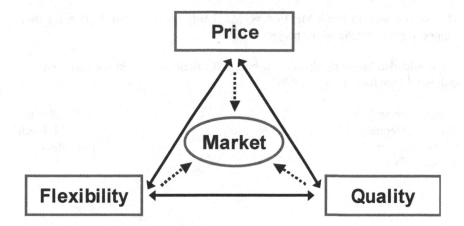

Fig. 2 Factors that influence market success

Sampling is just the beginning of the analytical process. On the way from sampling to the test report a lot of different requirements for high quality measurements have to be considered. There are external quality assurance requirements on the quality management system (e.g. accreditation, certification, GLP), internal quality assurance tools (e.g. method validation, the use of certified reference material, control charts) and external quality assurance measures (e.g. interlaboratory tests).

The aim of this book is to deal with all of these topics in a form that can easily be used for self-training and also for teaching in educational institutions and for in-house training. Teachers that intend to use this material to introduce the presented topics to their students or an audience are advised to study and digest the material before they use it in their presentations. The slides could then be customized to meet the needs of the teacher. It is important to note that the material provides the *basis* for presentations by third parties rather than exhaustive and fully comprehensive material.

The intention is to give an overview of all topics relevant for quality assurance in chemical measurement. For details on single topics we refer the reader to the relevant specialized literature. We have added some recent references for further studies and information at the end of each chapter.

The editors hope that they can contribute to a better understanding of quality assurance tools and the quality assurance system as a whole. They wish to promote the use of these tools in order to achieve world wide comparable measurement results.

The editors wish to thank Mr. Enders, Mr. Pauly and Springer–Verlag for their support throughout the whole project.

We would also like to thank all contributors for their work. Without their help this task would not have been possible.

Siegen, Germany   B. Wenclawiak
Stuttgart, Germany   M. Koch
Nicosia, Cyprus   E. Hadjicostas
Summer, 2010

List of Contributors

Michael Gluschke
Dottikon Exclusive Synthesis AG
Quality management
P.O. Box
5605 Dottikon
Switzerland
michael.gluschke@dottikon.com

Evsevios Hadjicostas
Quintessence Enterprises Ltd
Kennedy Business Center
Office 208
12-14 Kennedy Avenue
1087 Nicosia
Cyprus
quintessence@cytanet.com.cy

Rüdiger Kaus
Laboratory of Water Chemistry
    and Water Technology
University of Applied Science.
    Niederrhein
Adlerstr. 32
47798 Krefeld
Germany
rkaus@web.de

Michael Koch
Institute for Sanitary Engineering
University of Stuttgart
Bandtäle 2
70569 Stuttgart
Germany
michael.koch@iswa.uni-stuttgart.de

Bertil Magnusson
SP Technical Research
Institute of Sweden
Chemistry and Materials
    Technology
P.O. Box 857
50115Borås
Sweden
bertil.magnusson@sp.se

Ioannis Papadakis
International Quality Certification
Megistis 25
GR-17455, Alimos, Athens
Greece
papadakis@iqc.gr

Kyriacos C. Tsimillis
The Cyprus Organization for the
    Promotion of Quality –
    The Cyprus Accreditation Body
c/o Ministry of Commerce, Industry
    and Tourism
13-15 Andreas Araouzos str.
CY-1421, Nicosia, Cyprus
ktsimillis@cys.mcit.gov.cy

Bernd Wenclawiak
Inst. Analytische Chemie 1
Universität Siegen
FB 8 Chemie
Adolf-Reichwein-Str. 2
57078 Siegen
Germany
wenclawiak@chemie.uni-siegen.de

Contents

Important Information for Readers and Users of the Electronic Material

Viewing and Printing the Transparencies

The transparencies are available from the Springer Webserver under www.extras. springer.com/2010/978-3-642-13608-5.
You will find four zipped files there:

- Transparencies_English.zip
- Transparencies_German.zip
- ExcelKontrol_2.1_English.zip
- ExcelKontrol_2.1_German.zip

The first two contain all the transparencies in English and German language respectively and the latter two the control charts programme ExcelKontrol 2.1. To view the transparencies or use the programme contained in the zipped files, you will have to enter the password that you find printed at the end of Chap. 15. We strongly recommend that you download the zipped files to your own computer.

The transparencies are edited in Microsoft® PowerPoint® 2000. If you do not have Microsoft® PowerPoint® 2000 or a later version on your PC you can look at and print the transparencies with Microsoft® PowerPoint Viewer for PowerPoint, which is available free of charge on the Microsoft homepages.

Important Notice

After entering the password that you find printed at the end of Chap. 15, you may access the documents containing the transparencies via opening one of the files INDEX_DE.PPT or INDEX_EN.PPT in the respective folder and clicking on the hyperlinks, provided that you have extracted the zipped files completely on your own computer. Alternatively, you may click on the respective file names (*.ppt). When printing the overheads, please remember to set your printer to the right settings, regarding e.g. the medium of output (paper, overheads), colour, size. If necessary, please consult your PowerPoint® and/or printer handbook.

System Requirements

For use with PowerPoint® (recommended) the system requirements are described in the respective software manual.

EXCEL®-Files

The software EXCELKONTROL 2.1 is an EXCEL®-programme for control charts, for which Microsoft EXCEL 2000® (or later) is required.

Copyright and License

1. The transparencies in the book are protected by copyright. Any rights in them lie exclusively with Springer-Verlag, for EXCELKONROL the copyright is with the authors Dr. Michael Gluschke and Dr. Michael Koch.
2. The user may use the transparencies, print-outs thereof and multiple copies of the print-outs in classrooms and lecture halls. *All* copies most show the copyright notice of Springer-Verlag.
3. The user is entitled to use the data in the way described in section 2. Any other ways or possibilities of using the data are inadmissible, in particular any translation, reproduction, decompilation, transformation in a machinereadable language and public communication; this applies to all data as a whole and to any of their parts.

Liabilities of Springer-Verlag

1. Springer-Verlag will only be liable for damages, whatever the legal ground, in case of intent or gross negligence and with respect to warranted characteristics. A warranty of specific characteristics is given only in individual cases to a specific user and requires an explicit written representation. Liability under the product liability act is not affected hereby. Springer-Verlag may always claim a contributory fault on the part of the user.
2. The originator or manufacturer named on the product will only be liable to the user, whatever the legal ground, in case of intent or gross negligence.

ADDITIONAL CONDITIONS FOR USERS OUTSIDE THE EUROPEAN COMMUNITY: SPRINGER-VERLAG WILL NOT BE LIABLE FOR ANY DAMAGES, INCLUDING ANY LOST PROFITS, LOST SAVINGS, OR OTHER INCIDENTAL OR CONSEQUENTIAL DAMAGES ARISING FROM THE USE OF, OR INABILITY TO USE, THE ACCOMPANYING TRANSPARENCIES AND SOFTWARE, EVEN IF SPRINGER-VERLAG HAS BEEN ADVISED OF THE POSSIBILITY OF SUCH DAMAGES.

1 Glossary of Analytical Chemistry Terms (GAT)

Bernd Wenclawiak

Why is it so important to have a glossary of analytical terms? Because there are so many different acronyms, abbreviations, and incorrectly used 'terms', that even specialists sometimes have problems in understanding each other. A glossary is like a dictionary with the terms being the words in the vocabulary. Unfortunately not all words are found in one source. This chapter is a compilation of the most used terms.

Slide 1

Do you know all those terms on slide 9? Test yourself before you read the definitions given to you here in this chapter (and also in some of the other chapters). It is important, that each term at any time has the same meaning for every user.

Glossary

- Remark: teaching and learning terms is boring, but necessary!
- Why is it necessary to know the meaning of a term?
  Because a lot of expressions are not common in everyday life or might be interpreted differently
- In scientific and technical work it is necessary that all people use the same "language"

Slide 2

Here are some of the organisations and sources, which provide definitions on terms. Because many of them are often only referred to by their abbreviation, their full title is given here.
IUPAC - International Union of Pure and Applied Chemistry (www.iupac.org)
ISO - International Organization for Standardization (www.iso.ch)
IEC - International Electrotechnical Commission (www.iec.ch)

Glossary

- Who defines a term?
- In chemistry there is IUPAC
  IUPAC coordinates the international work of harmonization
- In the field of metrology (science of measurement) many different organizations work together:
  ISO, IEC , BIPM, OIML, IUPAC, IUPAP, IFCC
  They jointly published the VIM

B.W. Wenclawiak et al. (eds.), *Quality Assurance in Analytical Chemistry: Training and Teaching*, DOI 10.1007/978-3-642-13609-2_1, © Springer-Verlag Berlin Heidelberg 2010

BIPM - Bureau International des Poids et Measures (www.bipm.org)
OIML - International Organization of Legal Metrology (www.oiml.org)
IUPAP - International Union of Pure and Applied Physics (www.iupap.org)
IFCC - International Federation of Clinical Chemistry (www.ifcc.org/ifcc.asp)
VIM - International Vocabulary of Metrology - Basic and General Concepts and
associated Terms

Slide 3

In the sources mentioned in this slide
you will find more and in some cases
complete information on certain terms
and definitions. Many of the terms and
definitions given in this chapter are
taken from these sources.

Glossary - Vocabularies

- A Vocabulary contains general terms and definitions
  - International Vocabulary of Metrology – Basic and General Concepts and Associated Terms (VIM, 3rd edition, ISO/IEC-Guide 99:2007, www.bipm.org)
  - International Vocabulary of Terms in Legal Metrology (VIML)
  - Guide to the Expression of Uncertainty in Measurement (GUM, ISO/IEC-Guide 98:2008, www.bipm.org)
  - ISO 3534-1:2006 "Statistics - Vocabulary and symbols - Part 1: General statistical terms and terms used in probability"
  - ISO 3534-2:2006 "Statistics - Vocabulary and symbols - Part 2: Applied statistics "
  - ISO 3534-3:1999 "Statistics - Vocabulary and symbols - Part 3: Design of experiments "

Slide 4

Guides provide recommendations,
which are published by various
organisations.
Industrialisation lead to mass production with the characteristic feature of
the division of the work process into
smaller individual steps and the
simplest hand movements, which each
worker repeated incessantly. Different
parts only fit together if they are made
according to a standard. For example

Glossary

- The bibliography comprises four different types of documents:
  - Guides
  - Standards
  - Books
  - Articles (in scientific journals)

the inch is used as the unit for HPLC parts almost everywhere in the world, while
screws on the European continent have metric sizes and in the US the inch size is
still common.

Slide 5

ISO (International Organization of Standardization) was founded in Geneva in 1947

EN is European Norm

DIN (Deutsches Institut für Normung) is an Institute in Germany, which provides standardized industrial production/handling norms

BS or *BSI* (British Standards Institution) is British Standards, produced by the BSI

ASTM (formerly American Society for Testing and Materials) International is a global forum for the development of consensus standards organized in 1898, ASTM International is one of the largest voluntary standards developing organizations in the world. More than 12,000 ASTM standards can be found in the 80+-volume Annual Book of ASTM Standards.

Slide 6

The system of units used worldwide today is the International System of Units, in French, Système Inter national d'Unités (SI). The Bureau International des Poids et Mesures (BIPM) adopted the SI system at its 11th General Conference on Weights and Measures (Conférence Générale des Poids et Mesures –CGPM-) in 1960.

The mole was adopted as the seventh SI base unit in 1971. An important factor of the SI system of units is coherence, by which is meant that derived units are defined by the multiplication and/or division of the base units, without the need for any numerical factors.

Slide 7

Very important sources of information today are the websites on the Internet. *EURACHEM* (Co-operation for Analytical Chemistry in Europe) http://www.eurachem.org *EUROLAB* (Organization for Testing in Europe) http://www.eurolab.org *EA* (European Co-operation for Accreditation) http://www.european-accreditation.org/ *IRMM* (Institute for Reference Materials and Measurements; European Commission Joint Research Centre) http://irmm.jrc.ec.europa.eu. *CITAC* Cooperation on International Traceability in Analytical Chemistry http://www.citac.cc. *NIST* (agency of the US Commerce Department's Technology Administration) http://www.nist.gov. *ILAC* (International Laboratory Accreditation Cooperation) http://www.ilac.org. *APLAC* (Asia Pacific Laboratory Accreditation Cooperation) http://www.aplac.org/

Glossary - Important Organisations

- European
  - EURACHEM - Co-operation for Analytical Chemistry in Europe
  - EUROLAB - Organization for Testing in Europe
  - EA - European Co-Operation for Accreditation
  - EUROMET - A European Collaboration in Measurement Standards
  - IRMM Institute for Reference Materials and Measurements
- International
  - CITAC - Co-Operation on International Traceability in Analytical Chemistry
  - NIST – (US) National Institute of Standards and Technology
  - ILAC - the International Laboratory Accreditation Cooperation
  - APLAC Asia Pacific Laboratory Accreditation Cooperation

Slide 8

The terms presented here are separated into different fields for better clarity and to enable comparison. So you might want to follow up only those, of the six specific headings given here, that you need.

Glossary - Terms Related to:

- General
- Statistics
- Validation
- Measurement
- Error
- Uncertainty

Slide 9

This is the compilation of those terms, which can be related to topic "General". In the following slides the definitions will be given. In the upper left corner of the following slides you find the allocation to the main area the term is allocated to.

General Terms

Glossary of General Terms

- Accreditation
- Accuracy
- Accuracy of measurement
- Accuracy of a Measuring Instrument
- Audit
- Bias
- Certification
- Fitness for Purpose
- Influence quantity
- Precision
- Intermediate Precision

- Quality
- Quality Assurance
- Quality Control
- Internal Quality Control
- Standard
- Trueness
- Value
- Accepted Reference Value
- True Value
- Conventional True Value

Slide 10

The conformity assessment body could be a laboratory, the third-party could be the accreditor coming to your laboratory to inspect, whether the required documentation, manuals, procedures, or personnel are appropriate to perform the specific conformity assessment task e.g. determine PAHs by HPLC. If the laboratory (the conformity assessment body) and the personnel can do the job then accreditation might be granted. Being accredited can be of competitive advantage for laboratories. Sometimes contractors require the (analytical) work to be carried out in an accredited laboratory. Accreditation gives confidence to the customer that the laboratory will fulfil the requirements that are necessary for the work to be done competently. You find more about accreditation in chapter 2 of this book.

General Terms

Accreditation

- Third-party attestation related to a conformity assessment body conveying formal demonstration of its competence to carry out specific conformity assessment tasks [ISO/IEC 17000:2004]
  - (See also chapter 2)

Slide 11

Accuracy is the closeness of a result to a true value. This again is the combination of trueness and precision and defines measurement uncertainty. (See also chapter 12). Accuracy is greater when the quantity value is closer to the true value

General Terms

Accuracy

- Closeness of agreement between a measured quantity value and a true quantity value of a measurand [VIM]
  - Accuracy is a measure which combines precision and trueness (i.e. the effects of random and systematic factors)
  - Suppose the results produced by the application of a method show zero or very low bias (i.e. are "true"), their accuracy becomes equivalent to their precision. If the precision is poor, any particular result will be inaccurate
  - If a method shows a high bias, even results with a high precision are inaccurate
  - The concept 'measurement accuracy' is not a quantity and is not given a numerical quantity value [VIM].

Slide 12

For example: For contract work an auditor comes to the laboratory and checks whether the staff perform the task according to (agreed) standards, utilising appropriate laboratory equipment correctly. This is also called assessment or external audit. Internal auditors can be colleagues (from a different laboratory in the same company or a different working area). Reviews are usually carried out by upper level managers.
See also chapter 2.

General Terms

Audit

- Systematic, independent, documented process for obtaining records, statements of fact or other relevant information and assessing them objectively to determine the extent to which specified requirements (need or expectation that is stated) are fulfilled [ISO 17000]
  - Whilst "audit" applies to management systems, "assessment" applies to conformity assessment bodies as well as more generally.

Slide 13

Bias is the total systematic error (there may be more than one component contributing to total systematic error). It is the (positive or negative) differ-ence (Δ) of the population mean (μ, the limiting value of the arithmetic mean for n$\rightarrow\infty$) from the (known or assumed) true value (τ). $\Delta = \mu - \tau$. Therefore bias is the lack of trueness.

General Terms

Bias (Δ)

- The difference between the population mean (μ) and the 'true' value (τ) i.e., $\Delta = \mu - \tau$ (signed quantity) [IUPAC Orange book]
- Since the 'true' value is principally unknown, a conventional true value is used to estimate the bias

- Comment: Bias is the total systematic error

Slide 14

The confirmation of certain character-istics of a material, person, or organisation is called certification. This confirmation is often provided by an external audit or assessment (u.s.). Quality management systems are often certified for conformation with ISO 9000. Probably the most common term with respect to our topic here is 'certi-fied reference material' (CRM).
A driver's licence is a certificate,

General Terms

Certification

- Third-party attestation (issue of a statement) related to products, processes, systems or persons [ISO 17000]
  - Certification of a management system is sometimes also called registration.
  - Certification is applicable to all objects of conformity assessment except for conformity assessment bodies themselves, to which accreditation is applicable.

documenting the competence of the driver to participate in car traffic, usually obtained after an examination. (See also chapters 10 and 14)

Slide 15

Fitness for purpose is the ultimate goal of the person doing the job in the laboratory. It is also a requirement of the instrument used to perform an analysis and of the method chosen to get a correct result. Check in chapter 11 on this topic for further information.

General Terms

Fitness for Purpose

- Degree to which data produced by a measurement process enables a user to make technically and administratively correct decisions for a stated purpose. [IUPAC Orange Book]

Slide 16

Examples of influence quantities are:
- Temperature of a micrometer used to measure length;
- Frequency in the measurement of an alternating electric potential difference;
- Bilirubin concentration in the measurement of haemoglobin concentration in human blood plasma.

General Terms

Influence Quantity

- A **quantity** that, in a direct **measurement**, does not affect the quantity that is actually measured, but affects the relation between the **indication** and the **measurement result** [VIM]

Slide 17

If a number of results are close to each other we say that we have a good precision. Precision means proximity of test results. Keep in mind that they could be far away from the true value. In this case precision is still very good but the values are wrong on average. So the accuracy is pretty bad.

General Terms

Precision

- Closeness of agreement between **indications** or **measured quantity values** obtained by replicate **measurements** on the same or similar objects under specified conditions [VIM]

- The precision of a set of results of measurements can be quantified e.g. as a standard deviation

Slides 18-20

The quality of a product or a service is the degree to which a set of inherent characteristics fulfils requirements. For products this means e.g. properly made, defect free, made to agreed size. For service this means e.g. fast, reliable, and correct. Quality can be good or poor. The quality of data should always be as good as possible, but there is no need that they are better than required.

The word "quality" is derived from the Latin "qualitas", which means, incidentally, only the "nature" and "inherent characteristics" of a thing. In everyday speech we understand by this term above all two aspects, faultlessness ("a product with no defects" is of high quality) and performance capability or serviceableness ("a product that comes up to all our requirements and can be easily handled is qualitatively perfect").

See also chapter 6.

General Terms

Quality

- Degree to which a set of inherent characteristics fulfils requirements [ISO 9000]

- A very important remark:
 "No data is better than poor data"

General Terms

Quality - What does that Mean?

- "Quality means that the customer comes back, ...

General Terms

Quality - What does that Mean?

- ... and not the product".

Slide 21

Quality assurance is the main goal of this book. The authors want to provide the reader or user with a product or service (this book) that can satisfy the needs and which is hopefully fit for purpose.

General Terms

Quality Assurance

- Part of quality management focused on providing confidence that quality requirements will be fulfilled [ISO 9000]
- All the planned and systematic activities implemented within the quality system, and demonstrated as needed, to provide adequate confidence that an entity will fulfil requirements for quality [IUPAC Orange Book]

Slide 22

Everybody has heard or used that term. For example chromatography column manufacturers assure a certain specification e.g. the minimum plate number or separation efficiency for defined analytes. In our context that means: make sure that the instrument and our method works reliably within certain limits. To be certain that they really do, you should check this e.g. with a reference material (in liquid chromatography for example with an Engelhardt test solution). If there are deviations from previous performed tests, take action to correct this. In Analytical Chemistry control charts, analysis of certified reference materials and interlaboratory comparisons are very important quality control tools (see chapters 13-15).

General Terms

Quality Control

- Part of quality management (coordinated activities to direct and control an organization with regard to quality) focused on fulfilling quality requirements [ISO 9000]

Slide 23

You could decide to use control charts and to analyse certified reference materials (see chapter 13 and 14). Control charts are a simple, but effective tool for internal quality control. "Internal quality control is one of a number of concerted measures that analytical chemists can take to ensure that the data produced in the laboratory are fit for their intended purpose." (Cited from IUPAC Orange Book).

General Terms

Internal Quality Control IQC

- Set of procedures undertaken by laboratory staff for the continuous monitoring of operations and the results of measurements in order to decide whether results are reliable enough to be released [for complete definition see IUPAC Orange Book]

Slide 24

In this book, *standard* is used only in the sense of *written standard* and the term *measurement standard* or *etalon* (in French "étalon")(see slide 36) is used to describe chemical or physical standards used for calibration purposes such as: chemicals of established purity and their corresponding solutions of known concentration, UV filter, weights, etc. They are also called: reference materials.

General Terms

Standard

- The term standard has two different meanings:
  - A written standard is used for widely adopted procedures, specifications, technical recommendations etc.
  - And also as chemical or physical measurement standard, which is used for calibration purposes
- Standard is used here only in the sense of written standard

Slide 25

Trueness is a property related to systematic errors. It is the closeness of agreement between the average value obtained from a large set of test results and an accepted reference value. It can be checked with reference materials or in interlaboratory comparisons.

General Terms

Trueness

- Closeness of agreement between the average of an infinite number of replicate **measured quantity values** and a **reference quantity value** [VIM]
  - A reference quantity value is a value with little (or ideally no) systematical error
  - Perfect trueness cannot be achieved, so trueness in its analytical meaning is always trueness within certain limits
    - These limits may be narrow at a high concentration level and wide at the trace level
  - Note the difference between accuracy and trueness.
  - The lack of trueness is called bias

Slide 26

True value (of a quantity) is the ultimate goal when analyzing samples. However no measurement is perfect and thus true values are indeterminate by nature.

General Terms

True value (τ)

- Quantity value consistent with the definition of a quantity [VIM]
  - The true value is a theoretical concept and, in general, cannot be known exactly
  - It is a value that would be obtained by a perfect measurement
  - True values are by nature indeterminate

Slide 27

Values stated on reference materials are conventional true values. The results are usually obtained by independent methods in different expert laboratories. The conventional true value should be close enough to the true value.
In proficiency tests the mean of the participants results is often used as a conventional true value.

General Terms

Conventional True Value

- Value attributed to a particular quantity and accepted, sometimes by convention, as having an uncertainty appropriate for a given purpose [IUPAC Orange Book]
  - A result obtained by using several independent methods in several expert laboratories on one measurand is regarded as conventional true value of a quantity
    - even if it is not the "true" value
  - A conventional true value is in general, regarded as sufficiently close to the true value

Slide 28

This is a selection of terms related to statistics. You will find more detailed descriptions in chapter 8 – "Basic Statistics".

Glossary on Statistical Terms

- Arithmetic mean
- Distribution functions
- Normal distribution
- Rectangular distribution
- Triangular distribution
- Poisson distribution
- Binomial distribution
- Estimate

- Probabilty
- Standard Deviation
- Sample Standard Deviation
- Standard deviation of the mean
- Relative Standard Deviation (RSD)

Slide 29

Distributions are derived from data to give a mathematical description or a model for the data. You will find much more detailed information on this topic in textbooks on statistics.

Statistical Terms

Distribution Functions

- Normal distribution   continous distribution
- Rectangular distribution
- Triangular distribution

- Poisson distribution   discrete function
- Binomial distribution   discrete function

Slide 30

These are the terms that are associated with validation. Many of these terms are explained in detail in the chapters 11- "Fit for Purpose and Validation of Analytical Methods", and 8 - "Basic Statistics". So please check there. A few examples (of the more often used terms) are presented on the next slide.

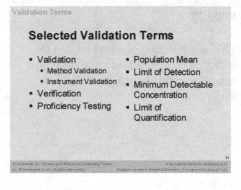

Slide 31

The ISO 9000:2005 states that *validation* is the confirmation by examination and provision of objective evidence that the particular requirements for a specified intended use are fulfilled.

Method Validation is the process of establishing the performance characteristics and limitations of a method and of verifying that a method is fit for purpose, i.e. for use for solving a particular analytical problem.

Instrument validation should show that an instrument is able to perform according to its design specification. This can be done for example by means of calibration or performance checks [Eurachem Guide Fit for Purpose]

Verification is confirmation by examination and provision of objective evidence that specified requirements have been fulfilled. ISO 9000:2005

Proficiency testing is a periodic assessment of the performance of individual laboratories and groups of laboratories that is achieved by the distribution by an independent testing body of typical materials for unsupervised analysis by the participants. [IUPAC Orange Book]

Population Mean (μ) The asymptotic value of the distribution that characterizes the measured quantity; the value that is approached as the number of observations approaches infinity. Modern statistical terminology labels this quantity the *expectation* or *expected value, E (x)*. [IUPAC Orange Book]

Limit of detection (LoD) is the lowest concentration that can be detected with specified confidence for a specific substance.

The *Minimum Detectable Concentration (MDC)* is the detection limit expressed as a concentration.

Limit of Quantification, (LoQ), sometimes also called limit of quantitation or limit of determination, is the minimum content that can be quantified with a certain confidence. Values below LoQ are reported as less than.

Slide 32

Calibration Curve is the graphical plot of the calibration function. The plot relates the signal to the analyte amount or concentration.
Linear least squares calibration is a calibration obtained using the method of minimizing the least squares. Sometimes calibration curves of higher order are used.

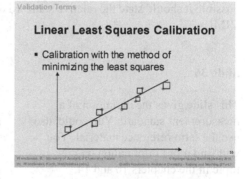

Slide 33

See chapter 9 – "Calibration and detection limits"

Slide 34

The *linearity* definition is on the slide. As long as the measurement result versus the analyte concentration fits a straight line we call this linearity. In an ideal case linearity could extend over several orders of magnitude. This is, for example, the case for certain elements when determined by ICP-OES. In other cases it can be less than one order of magnitude. Linearity over a large concentration can be very time saving, because fewer dilution steps might be necessary (see also chapter 9).

Slide 35

Measurand is the quantity we want to measure.

Measurement is the process of experimentally obtaining one or more quantity values that can reasonably be attributed to a quantity

Measurement procedure is the detailed description of a measurement. An operator should be able to measure a measurand following the description.

Measurement method is the generic description of a logical organization of operations used in a measurement

Measurement result is the information about the magnitude of a quantity, obtained experimentally. The information consists of a set of quantity values being attributed to the measurand together with any other available relevant information. This is usually summarized as a single quantity value and a measurement uncertainty. The single quantity value is an estimate, often an average or the median of the set. It should state the number of repetitions (n) used to obtain the averaged value and if possible it should state the relative standard deviation (RSD). [All definitions from VIM]

Glossary on Measurement Terms

- Measurand
- Measurement
- Measurement Procedure
- Measurement Method
- Measurement Result

Slide 36

This slide gives the definition of a measurement standard: You could also use the term reference material. You will find more information on this topic in the chapters 10 and 14.

The different definition of a (written) standard is on slide 24.

Measurement Standard

- Realization of the definition of a given quantity, with stated quantity value and associated measurement uncertainty, used as a reference [VIM]

Slide 37

Error of measurement can be due to random and systematic error.
Random Error is related to precision (see Slide 17) and Systematic Error to trueness (see Slide 25).

Error Terms

- Error of Measurement
- Random Error
- Systematic Error
- False Negatives/Positives

Slide 38

Error (of measurement) is the sum of random and systematic errors of one measurement. Since a true value cannot be determined, in practice a reference quantity value is used. Each individual result of a measurement will have its own associated error.

Error Terms

Error (of a Measurement)

- Measured quantity value minus a reference quantity value [VIM]
  - Error is the sum of random and systematic errors
  - If a measurement is repeated, each individual result will have its own associated error

Slide 39

IUPAC's Orange Book states two kinds of errors (really erroneous decisions): the error of the first kind ("type I", false positive), and the error of the second kind ("type II", false negative. The probability of the type I error is indicated by α; the probability for the type II error, by β. Default values recommended by IUPAC for α and β are 0.05, each.
False positives/negatives may be determined as follows:

Error Terms

False Positives / Negatives

- For qualitative methods the false positives/negatives rate may be determined.
- Data from a confirmatory method comparison should be provided if such method(s) is applicable to the same matrix(es) and concentration range(s).
- In the absence of a method comparison, populations of negative and positive fortified samples must be analysed.
  [Eurachem Fit for Purpose]

False positive rate (%) = false positives · 100/total known negatives
False negative rate (%) = false negatives · 100/total known positives

Slide 40

See also chapter 12 – "Measurement Uncertainty". Uncertainty is a fundamental property of a result.
Standard Uncertainty is the uncertainty of a measurement expressed as a standard deviation.
Combined Standard Uncertainty is the standard uncertainty that is obtained by combining (root of the sum of squares) individual standard measurement uncertainties associated with the input quantities in a measurement model.

Uncertainty

- Uncertainty (of Measurement)
  - Standard Uncertainty
  - Combined Standard Uncertainty
  - Expanded Uncertainty
  - Coverage Factor
  - Type A evaluation (of uncertainty)
  - Type B evaluation (of uncertainty)

Expanded Uncertainty is the combined standard uncertainty multiplied with the *coverage factor k*. Often k is chosen to be 2 or sometimes 3. With $k=2$ about 95%, and with $k=3$ about 99% of all likely values are encompassed.
Type A evaluation (of uncertainty): Method of evaluation of uncertainty by the statistical analysis of series of observations.
Type B evaluation (of uncertainty): Method of evaluation of uncertainty by means other than the statistical analysis of series of observations. [Definitions from VIM]

Slide 41

Uncertainty (of Measurement) i.e. Measurement Uncertainty. The slide contains the (newer) definition in VIM. In GUM it is defined as: "Parameter associated with the result of a measurement, that characterises the dispersion of the values that could reasonably be attributed to the measurand". The parameter may be, for example, a standard deviation (or a given multiple of it), or the half width of an interval having a stated level of confidence.

Measurement Uncertainty

- Non-negative parameter characterizing the dispersion of the quantity values being attributed to a measurand, based on the information used [VIM]
  - Uncertainty of measurement comprises, in general, many components. Some of these components may be evaluated from the statistical distribution of the results of series of measurements and can be characterized by experimental standard deviations
  - The other components, which can also be characterized by standard deviations, are evaluated from assumed probability distributions based on experience or other information

Slide 42

As stated on the slide, traceability should allow to trace the result of a measurement or the value of a standard through an unbroken chain of calibrations to stated references. Traceability is one key issue of ISO 17025. You find more on measurement traceability in chapter 10.

Metrological Traceability

• Property of a measurement result whereby the result can be related to a reference through a documented unbroken chain of calibrations, each contributing to the measurement uncertainty [VIM]

Bibliography

AOAC Terms and Definitions http://www.aoac.org/terms.htm

ISO (1998) SI-Guide, 32 p., ISBN 92-67-10279-6

ISO 3534-1:2006, Statistics - Vocabulary and symbols - Part 1: General statistical terms and terms used in probability

ISO 3534-2:2006, Statistics - Vocabulary and symbols - Part 2: Applied Statistics

ISO 3534-3:1999, Statistics - Vocabulary and symbols - Part 3: Design of experiments

ISO 9001:2000, Quality management systems – Requirements

ISO/IEC Guide 2:2004, Standardisation and related activities - General vocabulary

ISO/IEC Guide 98:1995 Guide to the expression of uncertainty in measurement (GUM), also available as JCGM 100:2008 from www.bipm.org

ISO/IEC Guide 99:2007 International Vocabulary of Metrology – Basic and General Concepts and Associated Terms (VIM), 3rd edition, also available as JCGM 200:2008 from www.bipm.org

IUPAC (1997) Gold book - Compendium of Chemical Terminology, http://goldbook.iupac.org/index.html

IUPAC (1998) Orange book- "Compendium on Analytical Nomenclature", 3rd edition, http://old.iupac.org/publications/analytical_compendium/

Neidhart B, Albus HE, Fleming J, Tausch C, Wegscheider W(1996-1998) Glossary of analytical terms. Accredit. Qual. Assur.

2   Accreditation – ISO/IEC 17025

Rüdiger Kaus

This chapter gives the background on the accreditation of testing and calibration laboratories according to ISO/IEC 17025 and sets out the requirements of this international standard. ISO 15189 describes similar requirements especially tailored for medical laboratories. Because of these similarities ISO 15189 is not separately mentioned throughout this lecture.

Slide 1

Chemical measurements are widely used for different purposes, e.g. as a basis for decision. Quite often the consequences of wrong decisions would be very costly. So the quality of chemical measurements is very important. Since customers of chemical laboratories are normally not able to judge the quality of a laboratory, an independent check of a laboratory by a third party is of utmost importance.

The Value of Chemical Measurements

- Depends much on the level of confidence that can be placed in the results
- Using quality assurance (QA) principles increases at least the likelihood of measurements being soundly based and fit for its purpose
- Since a usual customer is not able to judge the suitability of a QA system in a laboratory, a formal recognition by a competent third party increases customer's confidence

Slide 2

Accreditation is the confirmation of the competence of a testing or calibration laboratory by an independent third party, the accreditation body.
Normally laboratories are accredited for their fulfilment of the requirements described in the international standard ISO/IEC 17025.

What is Accreditation ?

- Third-party attestation related to a conformity assessment body (e.g. a testing laboratory) conveying formal demonstration of its competence to carry out specific conformity assessment tasks (e.g. testing) (ISO/IEC 17000:2004)

B.W. Wenclawiak et al. (eds.), *Quality Assurance in Analytical Chemistry: Training and Teaching*, DOI 10.1007/978-3-642-13609-2_2, © Springer-Verlag Berlin Heidelberg 2010

Slide 3

The structure of the accreditation
bodies in each country may be different
and the procedures for accreditation
can also vary to some extent. But
nevertheless they all have to fulfil the
requirements of ISO/IEC 17011:2004
"Conformity assessment -
General requirements for accreditation
bodies accrediting conformity assessment bodies". To further harmonize
these procedures and to guarantee
multilateral recognition international organisations have been set up by the
accreditation bodies, in Europe the European Co-operation for Accreditation (EA)
and worldwide the International Laboratory Accreditation Cooperation (ILAC).

Accreditation Body

- Usually there are national regulations for one
  or several accreditation bodies in each
  country
- There is cooperation between accreditation
  bodies in international organisations
  - Europe: European Accreditation Cooperation (EA)
  - Asia/Australia: Asian Pacific Laboratory
    Accreditation Cooperation (APLAC)
  - Worldwide: International Laboratory Accreditation
    Cooperation (ILAC)

Slide 4

This slide shows the members of an
International Multilateral agreement
between Accreditation bodies worldwide, which ensures the mutual
recognition of the accreditation for
testing and calibration results in the
cooperating countries.

**International Multilateral
Recognition Agreement (MRA)**

- Accreditation bodies from all over the world signed an
  International MRA on recognition and acceptance of
  test and calibration results
  - Africa - Egypt, South Africa, Tunisia
  - America - Argentina, Brazil, Canada, Costa Rica, Cuba,
    Guatemala, Mexico, USA
  - Asia/Australia - Australia, Hong Kong (China), PR China, India,
    Indonesia, Israel, Japan, Rep. Korea, Malaysia, New Zealand,
    Pakistan, Philippines, Singapore, Sri Lanka, Chinese Taipei,
    Thailand, Unit. Arab. Emirates, Vietnam
  - Europe - Austria, Belgium, Czech Republic, Denmark, Finland,
    France, Germany, Greece, Ireland, Italy, Netherlands, Norway,
    Poland, Portugal, Romania, Russian Fed., Slovakia, Slovenia,
    Spain, Sweden, Switzerland, Turkey, United Kingdom

Slide 5

These are the elements of an accreditation procedure. The first step for the
laboratory is to contact the accreditation body, which in response informs
the laboratory about the details of the
accreditation procedure. When the
contract between the laboratory and
the accreditation body is signed, assessors will be nominated. They will
carry out technical audits of the application document and on-site laboratory

Accreditation Procedure

- Application to the accreditation body
- Contract with the accreditation body
- Nomination and commissioning of assessors
- Technical audit of the application documents
- On-site laboratory assessment
- If necessary proficiency testing
- Assessment report
- Inspection of the report in the related
  committee
- Accreditation decision
- Publication

assessment. In many cases it is necessary for the laboratory to participate successfully in proficiency tests.

The assessment report of the assessors will be inspected by the members of a local
committee which makes a recommendation to accredit the laboratory (or not).

Slide 6

ISO/IEC 17025 is the basic standard
that is utilised by testing and calibration laboratories for implementing a
quality management system and they
are accredited for their implementation
of this standard. This standard contains
the general requirements for the
competence of testing and calibration
laboratories. It is one of the most
important standards for the worldwide
globalization of trade.

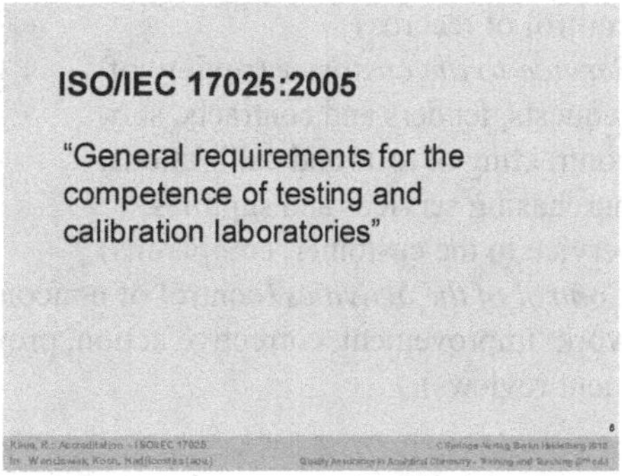

ISO/IEC 17025:2005

"General requirements for the
competence of testing and
calibration laboratories"

Slide 7

The standard ISO/IEC 17025 focuses
on the technical competence for specific tests and this is attested by the
accreditation body.

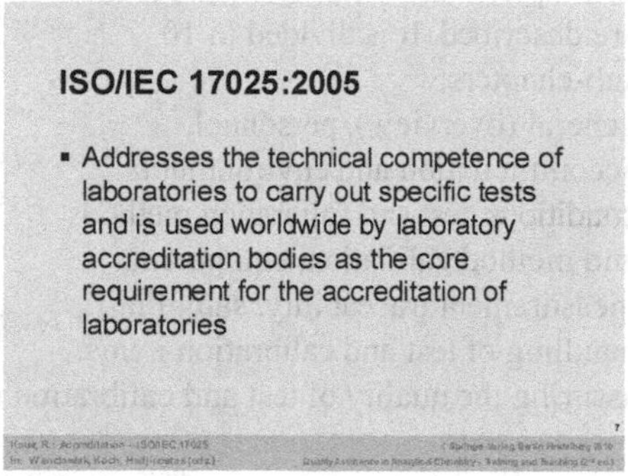

ISO/IEC 17025:2005

- Addresses the technical competence of
  laboratories to carry out specific tests
  and is used worldwide by laboratory
  accreditation bodies as the core
  requirement for the accreditation of
  laboratories

Slide 8

There is a prescribed form for international standards; this is followed by
ISO/IEC 17025, which has the general
chapters shown here at the beginning.

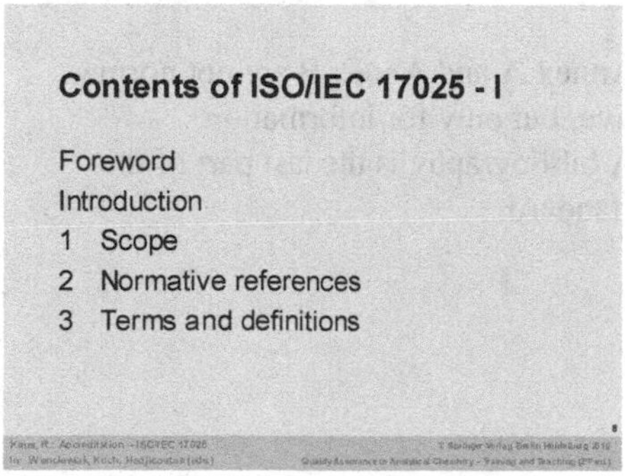

Contents of ISO/IEC 17025 - I

Foreword
Introduction
1   Scope
2   Normative references
3   Terms and definitions

Slide 9

Chapter 4 describes the management requirements. It is divided in 15 sub-chapters, which can be separated into three parts:
Organizational (organization, management system, document control, control of records)
Service to the customer (review of requests, tenders and contracts, subcontracting of tests and calibrations, purchasing services and supplies, service to the customer, complaints)
Control of the activities (control of nonconforming testing and/or calibration work, improvement, corrective action, preventive action, internal audits, management reviews)

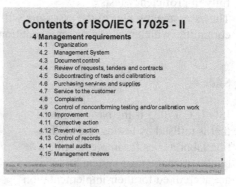

Slide 10

In chapter 5 the technical requirements are described. It is divided in 10 sub-chapters:
general (overview), personnel, accommodation and environmental conditions, test and calibration methods and method validation, equipment, measurement traceability, sampling, handling of test and calibration items, assuring the quality of test and calibration results, reporting the results.

Contents of ISO/IEC 17025 - III

5 Technical requirements
5.1 General
5.2 Personnel
5.3 Accommodation and environmental conditions
5.4 Test and calibration methods and method validation
5.5 Equipment
5.6 Measurement traceability
5.7 Sampling
5.8 Handling of test and calibration items
5.9 Assuring the quality of test and calibration results
5.10 Reporting the results

Slide 11

Annex A and Annex B are not normative, but only for information.
A bibliography is the last part of the standard.

Contents of ISO/IEC 17025 - IV

Annex A (informative)
Nominal cross-references to ISO 9001:2000

Annex B (informative)
Guidelines for establishing applications for specific fields

Bibliography

Slide 12

The formal statement of the activities is known as the "scope". Laboratories that wish to be accredited for their implementation of ISO/IEC 17025 have to apply quality assurance procedures to all or part of its operation. Where a laboratory claims compliance (e.g. GLP) against, or certification (e.g. ISO 9001) or accreditation (e.g. ISO/IEC 17025) to a particular standard, it is important to be clear to what this compliance, certification or accreditation applies.

Chapter 1 of ISO/IEC 17025: Scope

- A laboratory may apply quality assurance procedures according to ISO/IEC 17025 to all (or part) of its operations
- Where a laboratory claims compliance against or certification or accreditation to a particular standard, it is important to be clear to what this compliance, certification or accreditation applies
- The formal statement of the activities, which have been certified against ISO 9001, or accredited against ISO 17025 is known as the "scope"

Slide 13

The activities of the laboratory must be clearly stated.
The range of the work covered should be defined in detail, but the laboratory's operation should not be restricted by this definition.
Standards are often accused of hindering the development of methods.

Scope of ISO/IEC 17025 - I

- Quality management is supported by a clear statement of activities,
  - which ideally should define the range of work covered,
  - without restricting the laboratory's operation

Slide 14

Different quality standards have different rules, but for ISO/IEC 17025, which addresses laboratories, the scope may typically be defined in terms of:
- range of products, materials or sample types tested or analysed;
- measurements (or types of measurements) carried out;
- specification or method, equipment and technique used or
- concentration range and uncertainty as appropriate.

Scope of ISO/IEC 17025 - II

- is typically defined in terms of:
  - range of products, materials or sample types
  - measurements (or types of measurements)
  - specification or method/equipment/technique
  - concentration range and uncertainty

Slide 15

In the following slides the management requirements are described. This
chapter contains all elements of ISO
9001 which are relevant for testing and
calibration laboratories.
Therefore it is possible that a laboratory that undertakes the accreditation
procedure can get confirmation that it
additionally fulfils the requirements of
the ISO 9001.

Chapter 4 of ISO/IEC 17025

"Management requirements"

Slide 16

In chapter 4.1 of the standard the
requirements on the organization of the
laboratory are described.
The following themes are addressed:
legal responsibility, requirements of
this International Standard have to be
met, needs of the customers, of regulatory authorities or of organizations
providing recognition have to be
satisfied.
All this must be stated in the quality
policy of the laboratory (see 4.2.2 in ISO/IEC 17025).
The laboratory management system shall cover all the work carried out, not only
in the laboratory's permanent facilities, but also at sites away from its permanent
facilities or in associated temporary or mobile facilities.

Organization (Ch. 4.1)

- legal responsibility (4.1.1)
- meeting the requirements of this International Standard, satisfying the needs of the customer, the regulatory authorities or organizations providing recognition (4.1.2)
- The management system shall cover work carried out in all facilities:
  - laboratory's permanent facilities,
  - at sites away from its permanent facilities,
  - or in associated temporary or mobile facilities (4.1.3)

Slide 17

If the laboratory is part of an organization performing activities other than
testing and/or calibration, the responsibilities of the key personnel in the
organization that have an involvement
or influence on the testing and/or calibration activities of the laboratory
shall be defined in order to identify
potential conflicts of interest.

Organization (Ch. 4.1)

- Responsibilities of key personnel shall be defined in order to identify potential conflicts of interest (4.1.4)

Slide 18

In the following three slides the requirements for the organisation of the management of the laboratory are described in detail.

Organization (Ch. 4.1)

- Requirements on the laboratory (4.1.5) (1 of 3)
  - managerial and technical personnel with the authority and resources needed to carry out their duties ...
  - management and personnel free from any undue internal and external commercial, financial and other pressures and influences that may adversely affect the quality of their work
  - policies and procedures to ensure the protection of its customers' confidential information and proprietary rights...

Slide 19

Organization (Ch. 4.1)

- Requirements on the laboratory (4.1.5) (2 of 3)
  - policies and procedures to avoid involvement in any activities that would diminish confidence in its competence, impartiality judgement or operational integrity
  - define the organization and management structure of the laboratory, its place in any parent organization, and the relationships between quality management, technical operations and support services
  - specify the responsibility, authority and interrelationships of all personnel who manage, perform or verify work affecting the quality of the tests and/or calibrations

Slide 20

Organization (Ch. 4.1)

- Requirements on the laboratory (4.1.5) (3 of 3)
  - adequate supervision of testing and calibration staff, ...
  - technical management with overall responsibility for the technical operations ...
  - appoint a member of staff as quality manager ... with direct access to highest level of management
  - appoint deputies for key managerial personnel
    Individuals may have more than one function and it may be impractical to appoint deputies for every function

Slide 21

Chapter 4.2 demands that the laboratory must have a quality management
system appropriate to the scope of its
activities. This must be documented in
a quality manual which is written in an
understandable form and available to
all personnel (see chapter 7 in this
book).

Management System (Ch. 4.2)

- Establishing, implementing and maintaining a management system appropriate to the scope of its activities.
- Documenting its policies, systems, programmes, procedures and instructions to the extent necessary to assure the quality of the test and/or calibration results.
- The system's documentation shall be communicated to, understood by, available to, and implemented by the appropriate personnel. (4.2.1)

Slide 22

In the quality manual the laboratory's
management system policies and
objectives shall be defined. The
quality policy statement shall be
issued under the authority of the top
management. For more details on
quality manuals see chapter 7.

Management System (Ch. 4.2)

- Quality manual (4.2.2)
  - The laboratory's management system policies and objectives shall be defined in a quality manual (however named)
- Quality policy statement (4.2.2)
  - The overall objectives shall be documented in a quality policy statement. The quality policy statement shall be issued under the authority of top management

Slide 23

The laboratory's quality policy statement shall at least include the five
points listed in the next three slides.

Management System (Ch. 4.2)

- Contents of the quality policy statement (4.2.2) (1 of 3)
  It shall include at least the following:
  - the laboratory management's commitment to good professional practice and to the quality of its testing and calibration in servicing its customers
  - the management's statement of the laboratory's standard of service

Slide 24

> ## Management System (Ch. 4.2)
> - Contents of the quality policy statement (4.2.2)
>   (2 of 3)
>   - the purpose of the management system related to quality
>   - a requirement that all personnel concerned with testing and calibration activities within the laboratory familiarize themselves with the quality documentation and implement the policies and procedures in their work; and
>   - the laboratory management's commitment to compliance with this International Standard and to continually improve the effectiveness of the management system

Slide 25

> ## Management System (Ch. 4.2)
>
> - Contents of the quality policy statement (4.2.2)
>   (3 of 3)
>   - Note: The quality policy statement should be concise and may include the requirement that tests and/or calibrations shall always be carried out in accordance with stated methods and customers' requirements.
>   - Note: When the test and/or calibration laboratory is part of a larger organization, some quality policy elements may be in other documents.

Slide 26

In the predecessor standards and guides (EN 45001 and ISO-Guide 25) there are detailed descriptions of the contents of the quality manual. In ISO/IEC 17025 there are only some general statements about the contents of the manual.

> ## Quality Manual (Ch. 4.2.5 and 4.2.6)
>
> - The quality manual shall include or make reference to the supporting procedures including technical procedures.
> - It shall outline the structure of the documentation used in the management system.
> - The roles and responsibilities of technical management and the quality manager, including their responsibility for ensuring compliance with this International Standard, shall be defined in the quality manual.

Slide 27

The requirements on document control are described in chapter 4.3. These requirements are the most difficult parts of the standard for laboratories to implement.
In general these requirements are met by preparing an extensive master list.

Document Control (Ch. 4.3)

- General (4.3.1) (1 of 2)
  - The laboratory shall establish and maintain procedures to control all documents that form part of its management system ⇒ Master list (internally generated or from external sources)
  - such as regulations, standards, other normative documents, test and/or calibration methods, as well as drawings, software, specifications, instructions and manuals.

Slide 28

In a note the term "document" is specified.

Document Control (Ch. 4.3)

- General (4.3.1) (2 of 2)
  - In this context "document" could be policy statements, procedures, specifications, calibration tables, charts, text books, posters, notices, memoranda, software, drawings, plans, etc.
  - These may be on various media, whether hardcopy or electronic, and they may be digital, analog, photographic or written.
  - The control of data related to testing and calibration is covered in 5.4.7. The control of records is covered in 4.13.

Slide 29

The requirements or prerequisites of the Basis Documents are fixed. Basis documents have targets; they can be updated and are applicable over a longer time interval (e.g. standard operation procedures). Proof Documents contain proofs of operations and the details of the actual stock; they are not changeable and are only valid for single incidents (e.g. test reports).

Basis Documents ⇔ Proof Documents Differences

- Basis Documents
  (4.2.5 quality manual and 4.3 Document control)
  - Requirements or prerequisites
  - Target
  - Changeable
  - Valid for a longer period

- Proof Documents
  (4.13 Control of records)
  - Proof
  - Actual stock
  - Not changeable
  - Valid for a single incident

Slide 30

The review of the documents is a strenuous task. In chapter 4.3.2 the requirements for carrying out this review are listed.

Document Approval and Issue (Ch. 4.3.2)

- Review of the documents (4.3.2.1)
  - The procedure(s) adopted shall ensure that:
    - authorized editions of appropriate documents are available at all locations where operations essential to the effective functioning of the laboratory are performed;
    - documents are periodically reviewed and, where necessary, revised to ensure continuing suitability and compliance with applicable requirements;
    - invalid or obsolete documents are promptly removed from all points of issue or use, or otherwise assured against unintended use;
    - obsolete documents retained for either legal or knowledge preservation purposes are suitably marked. (4.3.2.2)

Slide 31

Unique identification is a key requirement for all QS documents.

Document Control (Ch. 4.3)

- Management system documents generated by the laboratory shall be uniquely identified (4.3.2.3)

Slide 32

The policies and procedures for the review of requests, tenders and contracts shall ensure that:
the requirements, including the methods to be used, are adequately defined, documented and understood the laboratory has the capability and resources to meet these requirements; an appropriate test and/or calibration method is selected that is capable of meeting the clients' requirements.

Review of Requests, Tenders and Contracts (Ch. 4.4)

- Policies and procedures (4.4.1)
- Records of reviews (4.4.2)
- Review for repetitive routine tasks

Any differences between the request or tender and the contract shall be resolved before any work commences. Each contract shall be acceptable both to the laboratory and the customer.

Records of reviews, including any significant changes, shall be maintained.

The requirements for the review of requests, tenders and contracts formulated in this standard are hard to fulfil in the routine work of a laboratory. For repetitive routine tasks, the review has to be made only at the initial enquiry stage or on the granting of the contract for on-going routine work performed under a general agreement with the customer, provided that the customer's requirements remain unchanged.

Slide 33

When a laboratory subcontracts work, either because of unforeseen reasons, or on a continuing basis, this work shall be placed with a competent sub-contractor. The subcontractor e.g. should comply with ISO/IEC 17025 for the work concerned. The laboratory is responsible to the customer for the subcontractor's work. The laboratory shall maintain a register of all subcontractors that are used for tests and/or calibrations and a record of the evidence of compliance with ISO/IEC 17025.

Subcontracting of Tests and Calibrations (4.5)

Purchasing Services and Supplies (4.6)

The laboratory shall have a policy and procedure(s) for the selection and purchasing of services and supplies it uses that affect the quality of the tests and/or calibrations. Procedures shall exist for the purchase, reception and storage of reagents and laboratory consumable materials relevant for the tests and calibrations. The laboratory shall evaluate suppliers of critical consumables, supplies and services that affect the quality of testing and calibration, and shall maintain records of these evaluations and a list of those approved.

Slide 34

Service to the customer is one of the key elements of this standard. The laboratory shall always have good contacts with its customers to clarify their requests and to monitor the laboratory's performance. Visits of the customers to the laboratory are welcome, provided that the laboratory is able to ensure the confidentiality of work for other customers.

Service to the Customer (Ch. 4.7)

- The laboratory shall be willing to cooperate with customers or their representatives in clarifying the customer's request and in monitoring the laboratory's performance in relation to the work performed,
- provided that the laboratory ensures confidentiality to other customers.

Slide 35

The laboratory shall have a policy and
procedure for the resolution of complaints received from clients or other
parties. Records shall be maintained of
all complaints and of the investigations
and corrective actions taken by the
laboratory. The laboratory shall have a
policy and procedures that shall be
implemented when any aspect of its
testing and/or calibration work, or the
results of this work, do not conform to

Ensuring the Quality

- Complaints (4.8)
- Control of nonconforming testing and/or
  calibration work (4.9)
- Improvement (4.10)
- Corrective action (4.11)
- Preventive action (4.12)

its own procedures or the agreed requirements of the client. The policy and procedures shall, among other things, ensure that the responsibilities and authorities for
the management of nonconforming work are designated and actions (including
halting of work and withholding of test reports and calibration certificates, as
necessary) are defined and taken when nonconforming work is identified.
In order to reduce the probability of non-conforming work, the laboratory shall
continuously improve its management system using appropriate policies, procedures, corrective actions, preventive actions and management reviews.
Where an evaluation indicates that the nonconforming work could recur or that
there is doubt about the compliance of the laboratory's operations with its own
policies and procedures, the corrective action procedures given in 4.11 shall be
promptly followed.
The laboratory shall establish a policy and procedure and shall designate appropriate
authorities for implementing corrective action when nonconforming work or
departures from the policies and procedures in the quality management system or
technical operations have been identified. The procedure for corrective action
shall start with an investigation determining the root cause(s) of the problem.
The laboratory shall monitor the results to ensure that the corrective actions taken
have been effective.
Where the identification of non-conformances or deviations generates doubts on
the laboratory's compliance with its own policies and procedures the laboratory
shall ensure that the appropriate areas of activity are audited in accordance with
4.14 (slide 37) as soon as possible.
Needed improvements and potential sources of non-conformances either technical
or concerning the quality management system, shall be identified. If preventive
action is required, an action plan shall be developed, implemented and monitored
to reduce the likelihood of the occurrence of such non-conformances and to take
advantage of the opportunities for improvement.

Slide 36

As already mentioned the standard distinguishes between documents and records. Records are proof documents. The laboratory shall establish and maintain procedures for handling such quality and technical records. Quality records shall include reports from internal audits and management reviews as well as records of corrective and preventive actions.

Control of Records (Ch. 4.13)

- The laboratory shall establish and maintain procedures for identification, collection, indexing, access, fixing, storage, maintenance and disposal of quality and technical records. (4.13.1.1)
- Quality records shall include reports from internal audits and management reviews as well as records of corrective and preventive actions.

Slide 37

Internal audits are very important elements in ensuring that the quality management system is always up to date. The laboratory shall periodically and in accordance with a predetermined schedule and procedure conduct internal audits of its activity to verify that its operations continue to comply with the requirements of the management system and ISO/IEC 17025.

Internal Audits (Ch. 4.14)

- The laboratory shall periodically and in accordance with a predetermined schedule and procedure
  - conduct internal audits of its activity
  - to verify that its operations continue comply with the requirements of the management system and this International Standard. ... (4.14.1)

Slide 38

In accordance with a predetermined schedule and procedure, the laboratory's executive management shall periodically conduct a review as described in the slide.

Management Reviews (Ch. 4.15)

- Laboratory's top management shall periodically conduct a review of the laboratory's management system and testing and/or calibration activities: ensure their continuing suitability effectiveness, and to introduce necessary changes or improvements. (4.15.1)

Slide 39

In chapter 4.15 the items of a management review are described in detail.

Management Reviews (Ch. 4.15)

- Review shall take into account (4.15.1):
  - the suitability of policies and procedures
  - reports from managerial and supervisory personnel
  - the outcome of recent internal audits
  - corrective and preventive actions
  - assessments by external bodies
  - the results of interlaboratory comparisons or proficiency tests
  - changes in the volume and type of the work
  - customer feedback
  - complaints
  - recommendations for improvement
  - other relevant factors, such as quality control activities, resources and staff training.

Slide 40

A typical period for conducting a
management review is once a year. It
is absolutely necessary to have feed
back from this review of the laboratory's work in order to profit from it.

Management Reviews (Ch. 4.15)

- A typical period for conducting a management review is once every 12 months.
- Results should feed into the laboratory planning system and should include the goals, objectives and action plans for the coming year.
- A management review includes consideration of related subjects at regular management meetings.
- Findings from management reviews and the actions that arise from them shall be recorded. The management shall ensure that those actions are carried out within an appropriate and agreed timescale. (4.15.2)

Slide 41

In chapter 5 of ISO/IEC 17025 the
"Technical requirements" are
described. These are detailed requirements for the special work in the
laboratories. They address testing
laboratories as well as calibration
laboratories. Therefore it is not always
easy to distinguish which requirement
applies only to calibration laboratories
and which to testing laboratories.

Chapter 5 of ISO/IEC 17025

"Technical Requirements"

Slide 42

In the general chapter the factors that determine the correctness and reliability of the tests and/or calibrations performed by a laboratory are listed together with the chapters in the standard where this aspect is covered.

General (Ch. 5.1)

- Factors which determine the correctness and reliability of the tests and/or calibrations performed by a laboratory (5.1.1)
  - human factors (5.2)
  - accommodation and environmental conditions (5.3)
  - test and calibration methods and method validation (5.4)
  - equipment (5.5)
  - measurement traceability (5.6)
  - sampling (5.7)
  - the handling of test and calibration items (5.8)

Slide 43

The estimation of the total uncertainty is a very important aspect in analytical measurements. In the predecessor version of the standard there was only one sentence about the statement of uncertainty ("on demand"). Now the estimation of uncertainty as well as the establishing of traceability plays a central role in the ISO/IEC 17025 (see chapters 10 and 12 of this book).
The laboratory shall take into account these factors in developing test and calibration methods and procedures, in the training and qualification of personnel, and in the selection and calibration of the equipment used.

General (Ch. 5.1)

- Influence to the total uncertainty (5.1.2)
  - The extent to which the factors contribute to the total uncertainty of measurement differs considerably between (types of) tests and between (types of) calibrations.
  - The laboratory shall take account of these factors in developing test and calibration methods and procedures, in the training and qualification of personnel, and in the selection and calibration of the equipment it uses.

Slide 44

The requirements for personnel are quite demanding and the assessors look in detail at the qualification of the personnel. The laboratory management shall ensure the competence of all who operate specific equipment, perform tests and/or calibrations, evaluate results, and sign test reports and calibration certificates. When staff are used which are still undergoing training, appropriate supervision shall be provided.

Personnel (Ch. 5.2)

- Requirements for personnel (5.2.1)
  - Ensuring the competence of all who operate specific equipment, perform tests and/or calibrations, evaluate results, and sign test reports and calibration certificates.
  - Personnel certification: the laboratory is responsible for fulfilling specified personnel certification requirements.

Slide 45

The laboratory shall have a training
programme for the personnel in which
the goals are formulated with respect
to the education, training and skills of
the laboratory personnel.
If contracted and additional technical
and key support personnel are used
(e.g. students), the laboratory shall
ensure that such personnel are super-
vised and competent and that they
work in accordance with the laboratory's management system.

Personnel (Ch. 5.2)

- Training programme (5.2.2)
- Supervising additional personnel (5.2.3)
- Authorization (5.2.5)

The management must authorize the personnel to perform specific tasks. The relevant authorization(s), competence, educational and professional qualifications,
training, skills and experience of all technical personnel, including contracted personnel shall be recorded. This can be done e.g. on an authorization sheet.

Slide 46

Accommodation and environmental
conditions in the laboratory shall be
such as to facilitate correct performance of the tests and/or calibrations.
The laboratory shall ensure that the
environmental conditions do not
invalidate the results or adversely
affect the required quality of any
measurement.
This shall be monitored (e.g. min-max
thermometer in refrigerators).

**Accommodation and
Environmental Conditions** (Ch. 5.3)

- Laboratory facilities (5.3.1)
- Monitoring the environmental conditions (5.3.2)
  - Tests and calibrations shall be stopped when the
    environmental conditions jeopardize the results of
    the tests and/or calibrations.
  - (Remember: "No data is better than poor data")
- Preventing cross-contamination (5.3.3)
- Access to the laboratory (5.3.4)
- Good housekeeping (5.3.5)

There must be an effective separation of rooms to prevent cross-contamination,
about which measures shall be taken. Particular care shall be taken when sampling
and tests and/or calibrations are undertaken at sites other than a permanent laboratory facility. The technical requirements for accommodation and environmental
conditions that can affect the results of tests and calibrations shall be documented.
The access to the laboratory shall be restricted to authorised personnel only. If
customers or other people visit the laboratories they must be accompanied. The
extent of control is based on the particular circumstances.
Measures shall be taken to ensure good housekeeping in the laboratory. Special
procedures shall be prepared where necessary, e.g. in microbiological laboratories.

Slide 47

The laboratory shall use appropriate methods and procedures for all tests and/or calibrations within its scope. These include sampling, handling, transport, storage and preparation of items to be tested and/or calibrated, and, where appropriate, an estimation of the measurement uncertainty (see chapter 12 of this book) as well as statistical techniques for analysis of test and/or calibration data.

Test and Calibration Methods and Method Validation (Ch. 5.4)

- General (5.4.1)
  - Using appropriate methods and procedures for all tests and/or calibrations within its scope.
  - The laboratory shall have instructions on the use and operation of all relevant equipment, and on the handling and preparation of items for testing and/or calibration,
  - shall be kept up to date and shall be made readily available to personnel

The laboratory shall have instructions on the use and operation of all relevant equipment, and on the handling and preparation of items for testing and/or calibration, or both, where the absence of such instructions could jeopardize the results of tests and/or calibrations. All instructions, standards, manuals and reference data relevant to the work of the laboratory shall be kept up to date and shall be made readily available to personnel (see also 4.3 in ISO/IEC 17025).

Deviations from test and calibration methods shall only occur if the deviation has been documented, technically justified, authorized, and accepted by the client.

Slide 48

The laboratory shall use methods, which meet the needs of the customer and which are appropriate for the tests and/or calibrations it undertakes. Standard methods (international, regional or national) shall preferably be used. The laboratory shall ensure that it uses the latest valid edition of a standard unless it is not appropriate or possible to do so.

Test and Calibration Methods and Method Validation (Ch. 5.4)

- Selection of methods (5.4.2)
- Laboratory developed methods (5.4.3)
- Non-standard methods (5.4.4)
- Validation of methods (5.4.5)
- Estimation of uncertainty of measurement (5.4.6)
- Control of data (5.4.7)

If no standard methods exist or if they are not appropriate, laboratory developed or other non-standard methods that are suitable for the special task and which are validated can be used. Validation of methods is an important task in the laboratory (see chapter 11).

The degree of rigour needed in estimating the uncertainty of measurement considerably differs between calibration and testing laboratories.

For testing methods it depends on factors such as: the requirements of the test method, the requirements of the customer, the existence of narrow limits on which decisions or conformance to a specification are based.

Calculations and data transfers shall be controlled in a systematic manner. Nowadays nearly all analytical measurements are made with computer-aided or automated equipment. The laboratory shall ensure the proper functioning to maintain the integrity of test and calibration data.

Commercial software (e.g. word processing, database and statistical programmes) may be considered to be sufficiently validated. However, there may be a need to validate laboratory software configuration/modifications.

Slide 49

The records of equipment shall include a lot of detailed information.

Equipment (Ch. 5.5)
Records of Equipment (5.5.5)

The records shall include at least the following:
* the identity of the item of equipment and its software
* the manufacturer's name, type identification, and serial number or other unique identification
* checks that equipment complies with the specification
* the current location, where appropriate
* the manufacturer's instructions, if available, or reference to their location
* dates, results and copies of reports and certificates of all calibrations, adjustments, acceptance criteria, and the due date of next calibration
* the maintenance plan, where appropriate, and maintenance carried out to date
* any damage, malfunction, modification or repair to the equipment

Slide 50

Establishing the traceability of the measurement is one of the most important requirements in ISO/IEC 17025. Measurement traceability ensures that the measurements in different laboratories are comparable in "space and time" all over the world. Calibration and the use of certified reference material (see chapters 9, 10 and 14) is the central tool in establishing traceability. Therefore a laboratory must have programmes and procedures for both.

Measurement Traceability (Ch. 5.6)

* The laboratory shall have an established programme and procedure for the calibration of its equipment.
* Such a programme should include a system for selecting, using, calibrating, checking, controlling and maintaining measurement standards, reference materials used as measurement standards, and measuring and test equipment used to perform tests and calibrations. ...

Slide 51

The programmes for calibration of equipment shall be designed and operated in a way to ensure that calibrations and measurements made by the calibration laboratory are traceable to the International System of Units (SI) (Système international d'unités). For testing laboratories, the extent to which the requirements in establishing traceability should be followed depends on the relative contribution of the calibration uncertainty to the total uncertainty. If calibration is the dominant factor, the requirements should be strictly followed.

Measurement Traceability (Ch. 5.6)

- Specific requirements (5.6.2)
  - for calibration laboratories
    - traceable to the International System of Units (SI) (Systeme international d'unites). (5.6.2.1.1)
  - Testing (5.6.2.2)
- Reference standards and reference materials (5.6.3)

Where traceability of measurements to SI units is not possible and/or not relevant, the same requirements for traceability to, for example, certified reference materials, agreed methods and/or consensus standards, are required as for calibration laboratories.

The laboratory shall have a programme and procedure for the calibration of its reference standards. A body that can provide traceability shall calibrate reference standards. Such reference standards of measurement held by the laboratory shall be used for calibration only and for no other purpose, unless it can be shown that their performance as reference standards would not be invalidated.

The laboratory shall have procedures for safe handling, transport, storage and use of reference standards and reference materials in order to prevent contamination or deterioration and in order to protect their integrity.

Slide 52

Sampling is a defined procedure whereby a part of a substance, material or product is taken as a representative sample of the whole for testing or calibration. The laboratory shall have sampling plans and procedures. They must be available at the location where sampling is undertaken. Sampling plans shall, whenever reasonable, be based on appropriate statistical methods. Any deviations, additions or exclusions from the documented sampling procedure shall be recorded in detail with the appropriate sampling data and shall be included in all documents containing

Further Chapters

- Sampling (Ch. 5.7)
- Handling of test and calibration items (Ch. 5.8)
- Assuring the quality of test and calibration results (Ch. 5.9)

test and/or calibration results, and shall be communicated to the appropriate personnel.

The laboratory shall have procedures for the handling (e.g. transportation, protection, storage, disposal) of test and/or calibration items.

To ensure the trackability (traceability to the raw data) of analytical results, the laboratory shall have a system to identify test and/or calibration items uniquely. The identification shall be retained throughout the life of the item in the laboratory.

All abnormalities or deviations from normal or specified conditions, as described in the test or calibration method, shall be recorded. When there is doubt on the suitability of an item for test or calibration, the laboratory shall consult the client for further instructions before proceeding and shall record the discussion.

The laboratory shall have quality control procedures for monitoring the validity of tests and calibrations undertaken. Trends should be detectable and, where practicable, statistical techniques shall be applied to the reviewing of the results (see chapter 12).

Slide 53

The results shall be reported in a test report or a calibration certificate, and shall include all the information requested by the customer and necessary for the interpretation of the test or calibration results and all information required by the method used.

In some cases (written agreement with the client) the results may be reported in a simplified way, but any information listed in 5.10.2 to 5.10.4 (see following slides) that is not reported to the client shall be readily available in the laboratory.

Reporting the Results (Ch. 5.10)

shall be reported
- accurately,
- clearly,
- unambiguously
- objectively
- in accordance with any specific instructions in the test or calibration methods.

Slide 54

In chapter 5.10.2 of ISO/IEC 17025 the form of the test reports and calibration certificates are specified in detail. The first points (title, name of laboratory and client, serial number, method) are necessary to guarantee unique identification.

Reporting the Results (Ch. 5.10)
Test Reports and Calibration Certificates
(5.10.2) (1 of 3)

Each test report or calibration certificate shall include at least the following information, unless the laboratory has valid reasons for not doing so:
- a title (e.g. "Test Report" or "Calibration Certificate")
- the name and address of the laboratory, and the location where the tests and/or calibrations were carried out, if different from the address of the laboratory
- unique identification of the test report or calibration certificate (such as the serial number), and on each page an identification in order to ensure that the page is recognized as a part of the test report or calibration certificate, and a clear identification of the end of the test report or calibration certificate
- the name and address of the customer
- identification of the method used

Slide 55

The following slides contain the details of the information that usually have to be included in the test or calibration certificate. These include the date of the measurement, the result, the staff responsible for carrying out the work and where relevant a statement that the results apply to the items tested or calibrated.

Reporting the Results (Ch. 5.10)
Test Reports and Calibration Certificates
(5.10.2) (2 of 3)

- a description of, the condition of, and unambiguous identification of the item(s) tested or calibrated
- the date of receipt of the test or calibration item(s) where this is critical to the validity and application of the results, and the date(s) of performance of the test or calibration
- reference to the sampling plan and procedures used by the laboratory or other bodies where these are relevant to the validity or application of the results
- the test or calibration results with, where appropriate, the units of measurement
- the name(s), function(s) and signature(s) or equivalent identification of person(s) authorizing the test report or calibration certificate
- where relevant, a statement to the effect that the results relate only to the items tested or calibrated

Slide 56

In a note to chapter 5.10.2 of ISO/IEC 17025 details of the certificates/reports are specified, including the need to number each page and to give the total number of pages.
A statement specifying that the test report or calibration certificate shall not be reproduced except in full, without written approval of the laboratory should be included in the document (e.g. as a footnote).

Reporting the Results (Ch. 5.10)
Test Reports and Calibration Certificates
(5.10.2) (3 of 3)

- Hard copies of test reports and calibration certificates should also include the page number and total number of pages.
- It is recommended that laboratories include a statement specifying that the test report or calibration certificate shall not be reproduced except in full, without written approval of the laboratory

Slide 57

In the preceding version of the standard it was strictly forbidden to give any interpretation of the results within the test report. Now it is explicitly allowed. In addition to the necessary requirements as listed above, test reports can, include the items listed in this slide.

Reporting the Results (Ch. 5.10)
Test Reports (5.10.3)

- In addition to the requirements listed in 5.10.2, test reports shall, where necessary for the interpretation of the test results, include the following:
  - deviations from, additions to, or exclusions from the test method, and information an specific test conditions, such as environmental conditions;
  - Where relevant, a statement of compliance/non-compliance with requirements and/or specifications; where applicable, a statement on the estimated uncertainty of measurement; information on uncertainty is needed in test reports when it is relevant to the validity or application of the test results, when a customer's instruction so requires, or when the uncertainty affects compliance to a specification limit;
  - where appropriate and needed, opinions and interpretations (see 5.10.5);
  - additional information which may be required by specific methods, customers or groups of customers. (5.10.3.1)

Slide 58

If test reports contain the results of
sampling they shall include additional
information about the sampling.

> **Reporting the Results** (Ch. 5.10)
> **Test Reports** (5.10.3)
> - the results of sampling shall include the following,
>   where necessary for the interpretation of test results:
>   - the date of sampling
>   - unambiguous identification of the substance, material or
>   product sampled (including the name of the manufacturer,
>   the model or type of designation and serial numbers as
>   appropriate)
>   - the location of sampling, including any diagrams, sketches or
>   photographs
>   - a reference to the sampling plan and procedures used
>   - details of any environmental conditions during sampling that
>   may affect the interpretation of the test results
>   - any standard or other specification for the sampling method
>   or procedure, and deviations, additions to or exclusions from
>   the specification concerned. (5.10.3.2)

Slide 59

In addition to the requirements listed
in clause 5.10.2 of ISO/IEC 17025,
calibration certificates shall include
conditions under which the calibration
was carried out, the uncertainty on the
result and information about the
measurement traceability, which is
necessary for the interpretation of
calibration results.

> **Reporting the Results** (Ch. 5.10)
> **Calibration Certificates** (5.10.4)
>
> - shall include:
>   - the conditions (e.g. environmental) under which
>   the calibrations were made that have an influence
>   an the measurement results
>   - the uncertainty of measurement and/or a
>   statement of compliance with an identified
>   metrological specification or clauses thereof
>   - evidence that the measurements are traceable

Slide 60

When opinions and interpretations are
included, the laboratory shall
document the basis upon which the
opinions and interpretations have been
made. Opinions and interpretations
shall be clearly marked as such in a
test report.

> **Reporting the Results** (Ch. 5.10)
> **Opinions and Interpretations** (5.10.5)
>
> - Documenting the basis upon which the
>   opinions and interpretations have been
>   made
> - Opinions and interpretations shall be
>   clearly marked as such in a test report.

Slide 61

Opinions and interpretations should not be confused with inspections and product certifications as intended in ISO/IEC 17020 and ISO/IEC Guide 65.
Opinions and interpretations included in a test report may comprise, but not be limited to those listed in the slide.

> **Reporting the Results (Ch. 5.10)**
> **Opinions and Interpretations (5.10.5)**
>
> - No confusion with inspections and product certifications as intended in ISO/IEC 17020 and ISO/IEC Guide 65 (5.10.5 Note 1)
> - Possible contents:
>   - an opinion on the statement of compliance/non-compliance of the results with requirements
>   - fulfilment of contractual requirements
>   - recommendations and how to use the results
>   - guidance to be used for improvements (5.10.5 Note 2)

Slide 62

When the test report contains results of tests performed by subcontractors, these results shall be clearly identified. Many laboratories have problems with this requirement, because they are afraid of the possibility that the clients would have doubt about their own competence or that the client could transfer their orders directly to sub-contractors.
In the case of transmission of test or calibration results by telephone, telex, facsimile or other electronic or electro-magnetic means, the requirements of this International Standard shall be met. The transmission must be done in a way that meets the requirements of the customer.

> **Reporting the Results (Ch. 5.10)**
>
> - Testing and calibration results obtained from subcontractors (5.10.6)
>   - Clear Identification
> - Electronic transmission of results (5.10.7)
>   - Requirements of this International Standard shall be met (see also 5.4.7)

Slide 63

The format shall be designed to accommodate each type of test or calibration carried out and to minimize the possibility of misunderstanding or misuse.
Attention should be given to the layout of the test report or calibration certifi-cate, especially with regard to the presentation of the test or calibration data and ease of assimilation by the reader. The headings should be standardized as far as possible.

> **Reporting the Results (Ch. 5.10)**
> **Format of Reports and Certificates (5.10.8)**
>
> - Accommodating each type of test or calibration carried out
> - Minimizing the possibility of misunderstanding or misuse.
> - Headings should be standardized as far as possible.

Slide 64

If there are amendments to test reports or calibration certificates after issue, they shall be made only in the form of a further document. It must contain a statement to which document the amendment belongs.
If there are too many corrections it is sometimes necessary to write a completely new test report or calibration certificate.
This new document shall be uniquely identified and shall contain a reference to the original that it replaces.

> **Reporting the Results (Ch. 5.10)**
> **Amendments to Test Reports and Calibration Certificates (5.10.9)**
>
> • Material amendments shall include the statement:
>   • "Supplement to Test Report [or Calibration Certificate], serial number ... [or as otherwise identified]", ...
> • When it is necessary to issue a complete new test report or calibration certificate, this shall be uniquely identified and shall contain a reference to the original that it replaces

Slide 65

As can be seen from the previous slides, preparation for accreditation is a tremendous amount of work. Nevertheless it will be increasingly necessary for laboratories, especially if they want to be active in different countries.
The accreditation according to ISO/IEC 17025 (or ISO 15189 for medical laboratories) has turned out to be the most important and worldwide valid standard for competent testing and calibration laboratories.

> **Summary**
>
> • Preparation for accreditation according to ISO/IEC 17025 is not an easy task, especially when done for the first time. But it assures at most customers confidence about quality and correct procedures.

Slide 66

More guidance on the way to accreditation may be found in the CITAC/EURACHEM guide mentioned here and the literature listed below.

> **More Guidance**
>
> • For laboratories preparing for accreditation according to ISO/IEC 17025 more guidance can be found in:
>
> CITAC/EURACHEM-Guide: Guide to Quality in Analytical Chemistry - An Aid to Accreditation, 2002, available from http://www.eurachem.org

Bibliography

CITAC/EURACHEM (2002) Guide to Quality in Analytical Chemistry – An Aid to Accreditation, available from www.eurachem.org

Günzler H (1996) Accreditation and Quality Assurance in Analytical Chemistry, Springer, Berlin

ISO/IEC 17000:2004 - Conformity assessment - Vocabulary and general principles

ISO/IEC 17011:2004 - Conformity assessment - General requirements for accreditation bodies accrediting conformity assessment bodies

ISO 17020:1998 - General criteria for the operation of various types of bodies performing inspection

ISO/IEC 17025:2005 - General requirements for the competence of testing and calibration laboratories

ISO 15189:2007 - Medical laboratories -- Particular requirements for quality and competence

ISO/IEC Guide 65:1996 - General requirements for bodies operating product certification systems

Wilson S, Weir G (1995) Food and Drink Laboratory Accreditation: A Practical Approach, Springer-Verlag, Berlin

3 ISO 9000 Quality Management System

Evsevios Hadjicostas

The ISO 9000 series describes a quality management system applicable to any organization. In this chapter we present the requirements of the standard in a way that is as close as possible to the needs of analytical laboratories. The sequence of the requirements follows that in the ISO 9001:2008 standard. In addition, the guidelines for performance improvement set out in the ISO 9004 are reviewed. Both standards should be used as a reference as well as the basis for further elaboration.

Slide 1

A laboratory product is the result, including intermediary results, of any laboratory process, e.g. the result of an analysis. A laboratory process is any activity employed for the production of the analysis results, e.g. the performance of an analysis etc. The laboratory customers are those having an interest in the analysis results. "The satisfaction" refers to the laboratory customers and interested parties as to the quality of both the laboratory products and laboratory processes. "The improvement" refers to the laboratory objective to achieve excellence through the continuous improvement of its products and processes.

Definitions

- The laboratory product
- The laboratory process
- The laboratory customer
- The satisfaction
- The improvement

B.W. Wenclawiak et al. (eds.), *Quality Assurance in Analytical Chemistry: Training and Teaching*, DOI 10.1007/978-3-642-13609-2_3, © Springer-Verlag Berlin Heidelberg 2010

Slide 2

A laboratory
- accepts inputs,
- adds value to these inputs and
- produces outputs for the customers.

Inputs are such as reagents, samples, methods, instruments, i.e. what is necessary for the proper operation of the laboratory. The output is what is supplied to the customer, i.e. the analysis results and expert opinion. Elements associated with the process are the following:
- people,
- equipment,
- materials and
- environment.

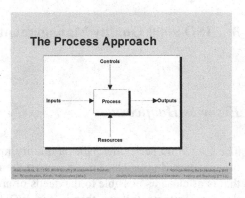

Slide 3

The requirements of the standard can be split into four major groups, which are those mentioned as "activities" in the quality management process. A fifth group is the quality management system per se. The model shown in the slide, taken from the ISO 9004: 2000 standard, illustrates the process link-ages as well as the interactions between the interested parties and the laboratory. Monitoring the satisfaction of interested parties requires the evaluation of information as to whether the laboratory has met their requirements.

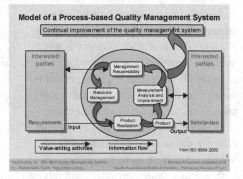

Slide 4

The first requirements are concerned with the quality management system (QMS) and the documentation that is necessary for the effective operation of the system, which is quite substantial. The numbering of the paragraphs is the same as in the standard to make it easier to look up further details.

4 Quality Management System

- 4.1 General requirements
- 4.2 Documentation requirements
  - 4.2.1 General
  - 4.2.2 Quality manual
  - 4.2.3 Control of documents
  - 4.2.4 Control of records

Slide 5

The laboratory shall establish, document, implement and maintain a QMS and continually improve its effectiveness. Laboratory processes are managed in a way that ensures the effectiveness of the QMS and the continual improvement of the laboratory operation. The laboratory processes and their sequence and interaction shall be defined and their effective operation shall be considered. Such effectiveness

4 QUALITY MANAGEMENT SYSTEM
4.1 General Requirements

- Quality management system (QMS)
- Continual improvement
- Sequence and interaction of processes
- Effective operation and control of processes
- Availability of resources and information
- Monitor, measure and analyze processes
- Achieve planned results
- Improve processes
- Outsourcing of processes

is ensured by the provision of the necessary resources and information. The effective operation of the laboratory processes is monitored on the basis of the achievement of the planned results. The output of this monitoring process is the improvement in the overall laboratory function.

The defined systems and processes of the laboratory QMS need to be simple so as to be clearly managed and understood. However, complicated processes often need a great deal of attention to simplify their operation.

Any laboratory processes that are outsourced shall be identified within the QMS and the laboratory shall ensure control over such processes. Such a control depends on the significance of the outsourced activities to the conformity of the result to defined requirements. The degree and extent of control exerted to outsourced activities is defined and documented to the QMS.

Slide 6

The laboratory management shall define the documentation and records needed to establish, implement and maintain the QMS and support an effective and efficient operation of the laboratory processes. A statement of the quality policy and quality objectives conveys the management commitment to quality and the customer oriented practices. The entire QMS is defined and documented in the quality manual, which is the top level document of the system. The quality manual is connected to the procedures, work instructions and other supporting documents that make the quality system operate. Records from all vital activities are used as evidence of the effective operation of the system.

> **4 QUALITY MANAGEMENT SYSTEM**
> **4.2 Documentation Requirements**
>
> 4.2.1 General
> - Quality policy
> - Quality objectives
> - Quality manual
> - Procedures
> - Work instructions
> - Records

Slide 7

The laboratory quality manual is a document addressing the quality management system requirements and shall
- refer to the scope of the QMS,
- make reference to the laboratory procedures and
- describe the interaction between the processes of the QMS.

The Quality Manual is the top level document of the laboratory and can be used to substantiate the compliance of the laboratory to the QMS requirements (see chapter 7).

> **4 QUALITY MANAGEMENT SYSTEM**
> **4.2 Documentation Requirements**
>
> 4.2.2 Quality manual
> - The laboratory quality manual shall include:
>   - The scope of the quality management system
>   - Links to the laboratory procedures
>   - Interactions between the processes of the QMS

Slide 8

The documents required by the QMS shall be controlled. Such control is necessary to approve, review and update the documents and to identify changes and amendments. Documents that are used for the operation of the QMS shall be available at points of use and understood by the personnel that uses this documentation. Documents of external origin must be identified and their distribution must

4 QUALITY MANAGEMENT SYSTEM
4.2 Documentation Requirements
4.2.3 Control of documents
- Approval of documents
- Review, update and re-approval of documents
- Identification of changes and current revisions
- Availability of documents at points of use
- Documents to be legible and readily identifiable
- Control of external documents
- Control of obsolete documents
- Achieves

be controlled. Obsolete documents must be prevented from unintended use and easily identified and controlled. The achieving system is considered as part, and thus, being under the control of the QMS.

Slide 9

Records are considered as a special type of documents and shall therefore be identifiable and properly controlled. The system for their storage, protection, retrieval, and for their retention time and disposition is documented in written procedures. The laboratory records provide objective evidence of conformity to requirements and of the effective operation of the QMS.

4 QUALITY MANAGEMENT SYSTEM
4.2 Documentation Requirements

4.2.4 Control of records
- Objective evidence of conformity to requirements
- Legible, readily identifiable, retrievable
- Documented procedure for the identification, storage, protection, retrieval, retention time and disposition of records

Slide 10

The introduction of the QMS is a significant investment for the organization and needs resources and effort to be disseminated throughout the organization. For this reason, management responsibility is the primary element of the QMS that needs to be defined. These are the elements that need to be

5 Management Responsibility
- 5.1 Management commitment
- 5.2 Customer focus
- 5.3 Quality policy
- 5.4 Planning
  - 5.4.1 Quality objectives
  - 5.4.2 Quality management system planning
- 5.5 Responsibility, authority and communication
  - 5.5.1 Responsibility and authority
  - 5.5.2 Management representative
  - 5.5.3 Internal communication
- 5.6 Management review
  - 5.6.1 General
  - 5.6.2 Review input
  - 5.6.3 Review output

addressed by the management. Details of each element and how they are addressed by the management are given below.

Slide 11

Leadership, commitment and active involvement of the top management are essential for developing and main-taining an effective and efficient QMS to achieve benefits for laboratory customers and interested parties. Top management and leaders establish unity of purpose and directions as well as the vision, policies and strategic objectives of the laboratory. They create and maintain the internal environment and urge people to become fully involved in achieving the quality objectives and satisfying customer's requirements. They define methods for measurement of the laboratory performance and plan for the future. They provide for the necessary resources and they review the operation of the QMS to continually improve its effectiveness.

5 MANAGEMENT RESPONSIBILITY
5.1 Management Commitment

- Addressing customer requirements
- Establishing the quality policy
- Ensuring that quality objectives are established
- Conducting management reviews
- Ensuring the availability of resources

Slide 12

The success of the laboratory depends on understanding and satisfying the current and future needs and expecta-tions of present and potential customers and other interested parties.
The interested parties of the laboratory are
- the customers (internal and external),
- people in the laboratory and the whole organization in general,
- the owners and investors,
- the suppliers and partners and
- the society and the environment.

5 MANAGEMENT RESPONSIBILITY
5.2 Customer Focus

- Satisfaction of the needs and expectation of the current and future customers
- Internal and external customers
- The interested parties

Slide 13

Quality policy is the commitment of management that the laboratory is seeking continuous improvement. This policy includes the vision and strategy of the organization and makes it clear to interested parties that it is determined to succeed.
The laboratory quality policy should be appropriate to the purpose of the laboratory and should include the commitment to quality and compliance with the system requirements. The quality policy shall provide a framework for establishing and reviewing quality objectives. It shall be communicated and understood within the organization and it is subject to systematic review for continuing suitability.

5 MANAGEMENT RESPONSIBILITY
5.3 Quality Policy

- Appropriate to the purpose of the laboratory
- Commitment to compliance with requirements
- How quality objectives are established and reviewed
- Communicated and understood within the laboratory
- Reviewed for continuing suitability

Slide 14

The laboratory should focus on defining the processes needed to meet effectively and efficiently the quality objectives and the requirements consistent with the strategy of the laboratory. The establishment of the quality planning is based on the inputs and needs for such a planning to be in place. The implementation of the quality planning is based on the outputs, i.e. the necessary resources needed for the plan to be realized.
The quality objectives are defined with the contribution of the interested parties, especially those that are bound to accomplish these objectives. They shall be measurable and consistent with the quality policy. Both the quality policy and quality objectives shall be communicated and understood within the laboratory and reviewed for continuing suitability.
The quality objectives are established at relevant functions and levels within the laboratory based on the feedback taken from relevant findings from management reviews, from current performance, from self-assessment etc.

5 MANAGEMENT RESPONSIBILITY
5.4 Planning

5.4.1 Quality objectives
- The quality objectives of the various functions and levels within the laboratory shall be established and be consistent with the quality policy and the laboratory strategy

Slide 15

Quality planning in a laboratory is the process needed to meet the laboratory's quality objectives. An effective and efficient planning consists of the inputs (strategies and objectives, customers' needs and expectation, evaluation of data etc.) and the outputs (needs for resources, needs for improvement, etc.).
Changes in the QMS must be effectively and efficiently managed so that the integrity of the system is maintained and it continues to improve and conform to the requirements.

5 MANAGEMENT RESPONSIBILITY
5.4 Planning

5.4.2 Quality management system planning
- Meet the general requirements for the quality management system (4.1)
- Management of change

Slide 16

Laboratory personnel should be given appropriate responsibility and authority to be motivated, committed and actively involved in contributing to the achievement of the quality objectives. The responsibilities and authorities of laboratory personnel shall be defined, communicated and agreed with interested individuals.

5 MANAGEMENT RESPONSIBILITY
5.5 Responsibility, Authority and Communication

5.5.1 Responsibility and authority
- Responsibilities and authorities of laboratory personnel should be defined and communicated within the laboratory

Slide 17

The management representative is the person appointed by the top management to implement the policy, monitor, evaluate and coordinate the QMS. The management representative shall be a member of the organization and such activity cannot be outsourced. He or she ensures that the processes needed for the proper operation of the QMS are established, implemented and maintained. He or she reports to the top management on the performance of the QMS. The management representative is the person that promotes the image of the laboratory to the internal as well as the external environment of the laboratory.

5 MANAGEMENT RESPONSIBILITY
5.5 Responsibility, Authority and Communication

5.5.2 Management representative
- Member of the organization
- Ensures the proper operation of the quality management system
- Reports to the top management
- Promotes the awareness of customer requirements throughout the organization

Slide 18

The laboratory top management establishes communication channels within
the laboratory departments that ensure
effective and efficient operation of the
laboratory. Activities that help
communication include for example,
meetings, information on notice
boards, a "laboratory journal", e-mails
and web sites, employee surveys and
suggestion schemes etc.

> **5 MANAGEMENT RESPONSIBILITY**
> **5.5 Responsibility, Authority and**
> **Communication**
>
> 5.5.3 Internal communication
> - Communication channels within the
>   laboratory
> - Communication as the stimulus for an
>   effective QMS

Slide 19

Management systematically reviews
the realization of processes and
evaluates the efficiency of the system
by conducting management reviews.
Management reviews are platforms for
the exchange of new ideas, with open
discussions being stimulated by the
leadership of top management. The
management reviews are conducted at
planned intervals and include assessment of opportunities for improvement
and the need for changes of the QMS.

> **5 MANAGEMENT RESPONSIBILITY**
> **5.6 Management Review**
>
> 5.6.1 General
> - Review of the system at planned
>   intervals
> - Assessment of opportunities for
>   improvement
> - Needs for changes
> - Records of management reviews

Slide 20

Management reviews are conducted
based on the "review inputs" and any
decisions and actions taken as a result
of the management review are the
"review output". The information
taken as input to the management
reviews could be the results from audits
and self-assessment of the laboratory,
feedback from customers and results
from benchmarking activities, the
laboratory activity performance, the
status of corrective and preventive action etc.

> **5 MANAGEMENT RESPONSIBILITY**
> **5.6 Management Review**
>
> 5.6.2 Review input
> - Results of audit and self-assessment
> - Feedback from customer and from
>   benchmarking activities
> - Performance of the laboratory activity
> - Any recommendations for improvement
>   (corrective and preventive actions,
>   suggestions from laboratory personnel)

Slide 21

The top management shall use the outputs of the management reviews as inputs to improvement of processes and resource needs. The decisions made in the management review meetings could be used to set up the laboratory objectives and form the laboratory strategy.

5 MANAGEMENT RESPONSIBILITY
5.6 Management Review

5.6.3 Review output
- Decisions and actions resulted from the management review meetings that will improve the effectiveness of the system and the quality of the laboratory product

Slide 22

Now that the management is committed to quality and organized to put the system in place, there is the need to provide the necessary resources including human resource, infrastructure, work environment, information and, of course, financial resources.

6 Resource Management

- 6.1 Provision of resources
- 6.2 Human resources
  - 6.2.1 General
  - 6.2.2 Competence, awareness and training
- 6.3 Infrastructure
- 6.4 Work environment

Slide 23

It is the responsibility of the laboratory management to ensure that the necessary resources essential to the implementation of strategy and laboratory objectives and the realization of customer satisfaction are identified and made available. Resources may be people, infrastructure, work environment, information, suppliers and partners, natural resources and financial resources. The provision of resources is supplied in relation to the opportunities and constraints that the laboratory faces. Amongst the resources are included tangible and intangible resources (such as intellectual properties), information management and technology, enhancement of competence via training and learning, use of natural resources etc.

6 RESOURCE MANAGEMENT
6.1 Provision of Resources

- Provision of the resources needed for:
  - The implementation and maintenance of the QMS
  - Realization of customer satisfaction

Slide 24

Laboratory personnel which affect the quality of the results as well as the conformity of such results to specified requirements shall be competent enough. The laboratory shall encourage the involvement and development of its people by providing ongoing training, by defining their responsibilities and authorities, by facilitating involvement in objective setting and decision making, by recognizing and rewarding etc.

6 RESOURCE MANAGEMENT
6.2 Human Resources

6.2.1 General
- Appropriate education, training, skills and experience of the laboratory personnel

Slide 25

The laboratory management shall determine the necessary competence of people working in the laboratory and ensure that it is available for the effective and efficient operation of the laboratory. Improvement of people's competence comes through awareness and training. The management of the laboratory should encourage people to develop through training and education programs, which are relevant to the needs of the laboratory. The effectiveness of training and education is evaluated and people are urged to strengthen the knowledge gained from training and education courses.

6 RESOURCE MANAGEMENT
6.2 Human Resources

6.2.2 Competence, awareness and training
- Determination of necessary competence of the laboratory personnel
- Provision of training
- Evaluation of the effectiveness of the action taken
- Awareness of personnel of the relevance and importance of their activities
- Maintenance of training records

Slide 26

Infrastructure includes resources such as buildings, equipment, supporting services, information and communication technology etc. The infrastructure shall be determined, provided and maintained so that the laboratory product conforms to specified requirements.

6 RESOURCE MANAGEMENT
6.3 Infrastructure

- Building and workspace
- Process equipment
- Supporting services (transportation, communication)
- Information systems

Slide 27

Work environment motivates and
enhances satisfaction and performance
of people. A suitable work environment
considers factors such as
- hygiene, cleanliness, absence of
  noise, vibration or pollution,
- heat, humidity, light, airflow,
- ergonomics, workplace location,
  and facilities for laboratory
  personnel
- safety rules,
- social interactions etc.

6 RESOURCE MANAGEMENT
6.4 Work Environment

- The work environment in the laboratory
  shall be determined and managed in a
  way to encourage people in achieving
  conformity to product requirements

Slide 28

The product realization is the main
part of the QMS activity. Product
realization in the case of a laboratory
is the process of performing the analysis.
First, the laboratory plans its activities
and then it determines the customer
related process, defines the analysis
requirements and develops the
communication channels with the
customers. The third stage is to design
and develop new products, i.e. new
methods and new analysis processes if necessary or if so required.

7 Product Realization

- 7.1 Planning of
  product realization
- 7.2 Customer-
  related processes
  - 7.2.1 Determination
    of requirements
    related to product
  - 7.2.2 Review of
    requirements related
    to product
  - 7.2.3 Customer
    communication
- 7.3 Design and development
  - 7.3.1 Design and development
    planning
  - 7.3.2 Design and development inputs
  - 7.3.3 Design and development output
  - 7.3.4 Design and development review
  - 7.3.5 Design and development
    verification
  - 7.3.6 Design and development
    validation
  - 7.3.7 Control of design and
    development changes

Slide 29

At the fourth stage the laboratory
develops the purchasing activities, i.e. it
defines the purchasing processes,
collects purchasing information and verifies the purchased products. The fifth
step is to define the system of performing the laboratory activity, validate such activities, keep records and
trace back to raw data. The laboratory
instruments must be calibrated and always be monitored for their proper operation.

7 Product Realization

- 7.4 Purchasing
  - 7.4.1 Purchasing
    processes
  - 7.4.2 Purchasing
    information
  - 7.4.3 Verification of
    purchased products
- 7.5 Product and service
  provision
  - 7.5.1 Control of
    production and service
    provisions
- 7.5.2 Validation of
  processes for production
  and service provisions
- 7.5.3 Identification and
  traceability
- 7.5.4 Customer property
- 7.5.5 Preservation of
  product
- 7.6 Control of
  monitoring and
  measuring devices

Slide 30

The laboratory should specify and
document the laboratory processes
(e.g. the analytical methods) and the
laboratory resources (instruments,
equipment etc.) that are necessary to
produce the laboratory product i.e. the
analysis results.
The quality plan defines the inputs and
outputs of any laboratory process. For
example, the quality plans refer to the
analytical methods, the instruments

> **7 PRODUCT REALIZATION**
> **7.1 Planning of Product Realization**
>
> • Quality objectives and product specification
> • Establishment of processes and documents
>   and provision of resources specific to the
>   product
> • Verification, validation, monitoring, inspection,
>   and test activities specific to the product and
> • Criteria for product acceptance
> • Records

and laboratory equipment that are used for the analysis, the characteristics of the
analysis results etc. All laboratory processes (e.g. analytical methods) and laboratory
products (e.g. analytical results) should be subject to verification and validation to
ensure that they are fit for the purpose. The acceptance criteria should be defined
and records should always be kept as evidence of meeting the requirements.

Slide 31

The laboratory shall ensure that there
is an effective and efficient communication with its customers so that the
analysis requirements are unambiguous
and well understood by the customer.
The laboratory shall provide for
customer satisfaction as to the apparent
as well as to the hidden customer
requirements. The analysis process
must always be performed based on
the relevant statutory and regulatory

> **7 PRODUCT REALIZATION**
> **7.2 Customer Related Processes**
>
> 7.2.1 Determination of requirements related
> to the product
> • Requirements specified by the customer
> • Requirements necessary for specified or
>   intended use
> • Statutory and regulatory requirements
> • Requirements determined by the
>   organization

requirements. Any requirements determined by the organization under which the
laboratory operates must also be considered.

Slide 32

Customer requirements are explicit and implicit. The laboratory shall review the requirements specified by the customer prior to commitment to undertake the job and ensure that it has the ability to meet the defined requirements. Anything that is discussed and agreed with the customer is confirmed and recorded. If any changes to agreed requirements occur, the laboratory shall ensure that the relevant documents are amended and that the relevant personnel are made aware of the changed requirements.

> **7 PRODUCT REALIZATION**
> **7.2 Customer Related Processes**
> 7.2.2 Review of requirements related to the product
> - Definition of product requirements
> - Resolution of contract or order requirements
> - Ability to meet the defined requirements
> - Records of the results of the review and action
> - Confirmation of customer requirements and re-confirmation in case of changes
> - Awareness of relevant personnel

Slide 33

Communication with customers is vital for the improvement of the laboratory performance regarding services as well as the quality of analytical results. Such communication keeps the relationship live and establishes the reputation of the laboratory.

> **7 PRODUCT REALIZATION**
> **7.2 Customer Related Processes**
> 7.2.3 Customer communication
> - The laboratory shall communicate with its customers to get information on the laboratory products and feedback from customers including customer complaints

Slide 34

The laboratory activity and its services to customers are designed and developed so as to respond effectively and efficiently to the needs and expectations of its customers and interested parties. Factors such as safety and health, testability, usability, user-friendliness, the environment and identified risks should be considered when designing the laboratory process. The laboratory shall plan and control the design and development (D&D) stages and shall define the verification and

> **7 PRODUCT REALIZATION**
> **7.3 Design and Development**
> 7.3.1 Design and development (D&D) planning
> - Design and development stages
> - Review, verification and validation
> - Responsibilities and authorities
> - Effective communication between D&D groups
> - Update of planning output

validation of each stage. Validation is the action taken to ensure that the final product (the output) satisfies specified requirements, whilst verification is the action taken to ensure that what is expected is what happens i.e. it is a comparison of the input with the output.

The responsibilities and authorities of personnel related to the design and development shall also be determined.

The different D&D groups are coordinated by the laboratory management, which ensures that communications are effective.

Slide 35

The laboratory shall identify the inputs that affect the design and development of the laboratory processes and facilitate their effective and efficient performance. Inputs are defined from external as well as internal needs and expectations. External inputs are drawn from customers, suppliers, users of the laboratory output, statutory and regulatory requirements, international or national standards, industry codes of practice etc. Internal inputs are drawn from policies and objectives, needs and expectations of people, technological development, past experience etc.

7 PRODUCT REALIZATION
7.3 Design and Development

7.3.2 Design and development inputs
- Functional and performance requirements
- Statutory and regulatory requirements
- Information from similar design
- Other relevant information

Slide 36

The design output of a laboratory process or laboratory service should include information to enable verification and validation to planned requirements. They should meet the input requirements and provide appropriate information to the interested parties due to use or have a benefit from these outputs. Examples of the output of design and development include data demonstrating the comparison of process inputs to process outputs, process specification, testing specifications, acceptance criteria, training requirements etc.

7 PRODUCT REALIZATION
7.3 Design and Development

7.3.3 Design and development output
- Meeting the input requirements
- Providing information for purchasing, production and servicing
- Defining acceptance criteria
- Providing parameters for the safe and proper use of the product

Slide 37

The design and development processes are systematically reviewed to ascertain that the progress goes as planned. Any problems and opportunities for improvement are identified and necessary action is taken and any changes are controlled.

> **7 PRODUCT REALIZATION**
> **7.3 Design and Development**
>
> 7.3.4 Design and development review
> - Systematic reviews at suitable stages
> - Evaluation of D&D results
> - Identification of problems
> - Proposal of necessary action

Slide 38

The verification of the D&D process ensures that the D&D output meets the D&D input requirements.
The validation of the D&D process is the evaluation of the output and ensures that the resulting product is capable of meeting the requirements for the specified application or intended use.
D&D changes shall be identified, reviewed, verified and validated. They should be approved before implementation and their effect on constituent parts and products already delivered should be considered.
Records of both verification and validation and review of changes of the design and development and any necessary action shall be maintained.

> **7 PRODUCT REALIZATION**
> **7.3 Design and Development**
>
> - 7.3.5 Design and development verification
> - 7.3.6 Design and development validation
> - 7.3.7 Control of design and development changes

Slide 39

Materials and services that are purchased by the laboratory should conform to specified requirements and satisfy the defined needs. The laboratory shall define the purchasing requirements and ensure that such requirements are met by the purchased product. The laboratory suppliers should be evaluated for their quality and reliability and should always be monitored regularly. The evaluation of

> **7 PRODUCT REALIZATION**
> **7.4 Purchasing**
>
> 7.4.1 Purchasing process
> - Definition of purchase requirements
> - Conformance of purchased product to purchase requirements
> - Control to the supplier and the purchased product
> - Criteria for selection, evaluation and re-evaluation of suppliers
> - Evaluation results to be recorded

the laboratory suppliers is based on pre-defined criteria set up by the laboratory. The purchasing processes should consider the quality, cost, performance and delivery of the purchased materials and services, the identification and traceability of the purchased materials, the supplier profile etc.

Slide 40

The laboratory shall define the characteristics of the materials, reagents and instrument to be purchased and evaluate the ability of the eligible suppliers to supply the required materials and/or services. The verification process that is, the process to ensure that the purchased products conform to the defined requirements shall be defined and the necessary resources, including personnel, shall be provided.

7 PRODUCT REALIZATION
7.4 Purchasing

7.4.2 Purchasing information
- Definition of the characteristics of the product to be purchased
- What is required for the approval of the purchased product, procedures, processes and equipment?
- What are the necessary qualifications of personnel?
- What are the QMS requirements?

The laboratory suppliers should be considered as part of the laboratory's quality management system. The type and extend of control applied to the suppliers and the purchased materials shall be dependent upon the effect of the purchased product on the subsequent output.

Slide 41

Based on the verification process already defined, upon receipt of purchased material the laboratory inspects and verifies that the received materials conform to the purchase requirements. When the verification is made at the supplier's premises there is a need to make arrangements for such verification and for a definition of the method for the release of products.

7 PRODUCT REALIZATION
7.4 Purchasing

7.4.3 Verification of purchased product
- Inspection of purchased material
- Purchased product to meet specified purchase requirements
- Verification at the supplier's premises
  - Define verification arrangements
  - Define method of product release

Slide 42

This clause of the standard refers to the whole laboratory activity and defines the requirements to be carried out by the laboratory. In order to control the laboratory activity there is a need for
- the information relevant to the characteristics of the laboratory analysis to be available and
- the information relevant for the effective operation of the analytical method to be also available.

7 PRODUCT REALIZATION
7.5 Production and Service Provision

7.5.1 Control of production and service provision
- Production and service provision under controlled conditions such as
  - Availability of information relevant to the characteristics of the products
  - Availability of work instructions
- ...

Moreover, the laboratory operation should be carried out using relevant operating instructions and procedures.

Slide 43

7 PRODUCT REALIZATION
7.5 Production and Service Provision

In addition, suitable equipment and suitable monitoring and measuring devices should be available so that the laboratory processes operate effectively. The procedures for the release of the laboratory product, i.e. the acceptance of the laboratory results as well as that for the delivery of such results to the customers, have to be specified and controlled.

7.5.1 Control of production and service provision (contd.)
- Availability of suitable equipment
- Availability and use of monitoring and measuring devices
- Monitoring and measurement
- Release, delivery and post-delivery activities

Slide 44

7 PRODUCT REALIZATION
7.5 Production and Service Provision

The analytical methods and laboratory processes shall be validated when the resulting output cannot be verified. This validation shall demonstrate the ability of the methods and processes to achieve the planned results. This is verified by inspection or measurements that the process output (e.g. the laboratory results) is reasonable and

7.5.2 Validation of processes for production and service provision
- Ability of processes to achieve planned results
- Verification of processes by monitoring or measurement of the resulting output
- Validation of processes for production and service provision
- ...

within expectations. Analytical methods and laboratory processes shall be validated so that to ensure fitness for the purpose (see chapter 11).

Slide 45

The laboratory ensures that the processes that are used for the performance of the analysis operate within defined limits and that the equipment used for the performance of such processes are approved and the personnel using this equipment are qualified. The operation of the laboratory activity is based on the use of specific methods and procedures. Objective evidence that the whole laboratory activity is under control is supported by the relevant records that the laboratory keeps and manages. Processes shall be re-validated from time to time as well as after the occurrence of any changes that might affect the process.

7 PRODUCT REALIZATION
7.5 Production and Service Provision

7.5.2 Validation of processes for production and service provision (contd.)

- Arrangements for processes for production and service provision
  - Criteria for review and approval of the processes
  - Approval of equipment and qualification of personnel
  - Use of specific methods and procedures
  - Requirements for records
  - Re-validation

Slide 46

The laboratory shall establish a system to identify the laboratory results and data, and develop a system to trace back to the raw data in order to support its findings. For example, the laboratory reports, the laboratory log books as well as the retention of samples etc. must bear a characteristic number or a well defined code. The status of the laboratory product, i.e. the stage or progress of the work done should be identified.
The need for identification and traceability may arise from relevant statutory and regulatory requirements, benchmarking performance, contract requirements etc.

7 PRODUCT REALIZATION
7.5 Production and Service Provision

7.5.3 Identification and traceability

- Identification of the laboratory product (i.e. the laboratory results)
- Status of the laboratory product
- Traceability of laboratory results back to the raw data

Slide 47

Any property that belongs to the customer and is under the control of the laboratory in order to perform an operation or offer a service to this customer is administered with care. Customer property can include intellectual property. Examples of customer's

7 PRODUCT REALIZATION
7.5 Production and Service Provision

7.5.4 Customer property
- Protection of the customer's property that is under the control or is being used by the laboratory
- Customers to be informed if their product is damaged or unsuitable
- Records of any damaged or unsuitable property
- Intellectual property

property that is under the control or being used by the laboratory might be the following:
- reference materials,
- specifications, drawings, analytical methods or any other intellectual property.

Slide 48

The laboratory product should be protected from damage or misuse during internal processing and final delivery to customers.

> **7 PRODUCT REALIZATION**
> **7.5 Production and Service Provision**
>
> 7.5.5 Preservation of product
> - Processes for handling, packaging, storage, preservation and delivery of product
> - Prevention of damage, deterioration or misuse
> - Involvement of suppliers to effective protection of purchased materials

Slide 49

The devices and analytical instruments that are used in the laboratory to monitor and measure various parameters during the laboratory activity shall be controlled in a way that ensures that they are consistent with the monitoring and measurement requirements. The laboratory shall determine the monitoring and measurements to be undertaken, establish processes to ensure that this activity can be carried

> **7 PRODUCT REALIZATION**
> **7.6 Control of Measuring and Monitoring Devices**
>
> - Determination of the necessary monitoring and measurements
> - Processes to ensure that this activity is carried out
> - Determination of the necessary monitoring and measurement devices
>   ...

out effectively and determine the necessary monitoring and measurement devices. Monitoring and measuring processes in the laboratory include methods and devices for verification and validation of products and processes.

Slide 50

Measuring equipment shall be calibrated or verified at specified intervals
or prior to use and adjusted as necessary. The calibration status of each
piece of equipment shall be identified
and records of the calibration and
verification results shall be maintained.
When a piece of equipment is found
not to conform to requirements, the
previous measuring results shall be
corrected.

7 PRODUCT REALIZATION
7.6 Control of Measuring and
Monitoring Devices
- Calibration or verification of devices
- Adjustment of devices
- Identification of calibration status
- Protection from damage and deterioration
- Assessment of the validity of previous results
- Appropriate action to affected product or equipment
- Calibration records

Slide 51

The primary objective of the quality
management system is the continual
improvement of the management
processes and the development of the
organization via the satisfaction of the
interested parties in the organization.
The improvement of the management
of processes comes from the measurements of the performance and effectiveness of processes and the analysis of
the measurement results. This is the
last but very important element of the standard and it is analysed below.

8 Measurement, Analysis and
Improvement
- 8.1 General
- 8.2 Monitoring and measurement
  - 8.2.1 Customer satisfaction
  - 8.2.2 Internal audit
  - 8.2.3 Monitoring and measurement of processes
  - 8.2.4 Monitoring and measurement of product
- 8.3 Control of non-conforming product
- 8.4 Analysis of data
- 8.5 Improvement
  - 8.5.1 Continual improvement
  - 8.5.2 Corrective action
  - 8.5.3 Preventive action

Slide 52

The monitoring of the laboratory performance is required to demonstrate
that the laboratory product conforms
to the specified requirements and that
the laboratory activity conforms to the
quality management system requirements. Moreover, there is a need and it
is a requirement that the laboratory
continually improves the effectiveness
of the quality management system.

8 MEASUREMENT, ANALYSIS AND
IMPROVEMENT
8.1 General

- Implementation of processes for
  monitoring, measurement, analysis and
  improvement of the laboratory activity
  and the continual improvement of the
  effectiveness of the Quality
  Management System

Slide 53

The laboratory shall plan and establish processes to listen to the "voice of customer and other interested parties" and cooperate with them in order to anticipate future needs. Such information should be collected, analysed and used for improving the laboratory performance. Measurements of customer satisfaction as well as personnel satisfaction (within the laboratory) are vital for the evaluation of laboratory performance.

8 MEASUREMENT, ANALYSIS AND IMPROVEMENT
8.2 Monitoring and Measurement

8.2.1 Customer satisfaction
- Meeting customer requirements
- Processes to collect, analyze and use of customer-related information
- Customer surveys
- Measurement of customer satisfaction
- Customer complaints

Slide 54

The laboratory should ensure the establishment of an effective and efficient internal audit process to assess the strengths and weaknesses of the quality management system. The internal audit process provides an independent tool for obtaining objective evidence that the existing requirements have been met, since the internal audit evaluates the effectiveness and efficiency of the laboratory.

8 MEASUREMENT, ANALYSIS AND IMPROVEMENT
8.2 Monitoring and Measurement

8.2.2 Internal audit
- Internal Audit at planned intervals
- System conformance to planned arrangements and to the QMS
- Effective implementation and maintenance of the system

Slide 55

The internal audit program takes into consideration the status and importance of processes and areas to be audited. The results from previous audits and the special needs are also taken into account.
The internal audit is performed by qualified internal auditors who are members of the organization. However, auditors should not audit their own work.

8 MEASUREMENT, ANALYSIS AND IMPROVEMENT
8.2 Monitoring and Measurement

8.2.2 Internal audit
- Audit program
  - Status and importance of processes
  - Areas to be audited
  - Results of previous audit
- Selection of auditors
  - Objectivity and impartiality
  - Not to audit their own work

During the audit process the auditors inspect the effective and efficient implement-ation of the laboratory processes, the capability of processes, the performance results, the improvement activities and they are also looking for opportunities for improvement. The aim of the internal audit is to determine whether the quality management system conforms to the planned arrangements and if it is effectively implemented.

Slide 56

The internal audit procedure is documented and it has to be under-stood both by the auditors and those being audited. The objective of the internal audit is to contribute to the improvement of the organization via the elimination
-   of the non-conformities and
-   of the causes of such non-conformities.

In response to the internal audit results the laboratory takes improvement action in due time.

> **8 MEASUREMENT, ANALYSIS AND IMPROVEMENT**
> **8.2 Monitoring and Measurement**
> 8.2.2 Internal audit
> ▪ Internal audit procedure
> ▪ Elimination of non conformities and their causes
> ▪ Follow up activities

Slide 57

The performance of the laboratory processes as well as their output is measured and their ability to conform to requirements is monitored. When planned results are not achieved, corrective action shall be taken as appropriate to ensure conformity to the requirements. The decision whether a laboratory process or a laboratory result (the output) conforms to requirements is based on the accept-

> **8 MEASUREMENT, ANALYSIS AND IMPROVEMENT**
> **8.2 Monitoring and Measurement**
> 8.2.3 Monitoring and measurement of processes
> ▪ Ability of processes to achieve planned results
> ▪ Evaluation of process performance

ance criteria that the laboratory defined. Such acceptance criteria could be for example the measurement uncertainty or the response of a measuring instrument. The decision process that is used for ensuring that a laboratory result conforms to the specified requirements considers the analytical methods employed, the parameters to be measured, the equipment used for the analysis etc.

Slide 58

The laboratory shall ensure that its results conform to specified requirements. This can be done at appropriate stages of the analysis process.
The laboratory product is released and/or a laboratory report is approved by qualified personnel after it has been ensured that they conform to the specified requirements.

8 MEASUREMENT, ANALYSIS AND IMPROVEMENT
8.2 Monitoring and Measurement

8.2.4 Monitoring and measurement of product
- Ability of product to satisfy the purpose
- Conformity to the acceptance criteria
- Qualified persons to release products

Slide 59

Non-conformities in the laboratory could be, for example, any results that are considered as correct when they are not, or considered as incorrect when they are correct. Non-conformities in the laboratory could also be any process output that is not conforming to the input requirements. The laboratory should effectively and efficiently control the non-conformities in order to
- correct and
- prevent failures.

Non-conformities are corrected and relevant records are collected and analyzed. Such data can provide valuable information for improving the effectiveness and efficiency of the laboratory processes.

8 MEASUREMENT, ANALYSIS AND IMPROVEMENT
8.3 Control of non-conforming products
- Identification of non-conforming products
- Procedure for non-conforming products
- Elimination of non-conformity
- Release under concession by a relevant authority or by the customer
- Records for non-conformities and concession
- Re-verification of corrected products
- Completion of planned arrangements before release

Slide 60

The data obtained from measurements are analysed and the subsequent information is used for fact-based decision-making. Such data are used to demonstrate the suitability and effectiveness of the quality management system and to evaluate the continual improvement. The analysis of the data provides information relating to

8 MEASUREMENT, ANALYSIS AND IMPROVEMENT
8.4 Analysis of Data
- Analysis of data
  - Suitability and effectiveness of the QMS
  - Continual improvement
- Customer satisfaction
- Conformity to product requirements
- Trends of processes
- Opportunities for preventive action
- Performance of suppliers

customer satisfaction and the conformity of processes and products to the requirements.

For the analysis of data the laboratory shall employ appropriate statistical techniques. Analysis of the data can help to determine the root cause of existing or potential problems, and therefore indicate the corrective and preventive actions needed for improvement.

Slide 61

The laboratory shall continually improve the effectiveness of the quality management system through the use of the quality policy, quality objectives, audit results, analysis of data, corrective and preventive actions, and management review. Improvements can range from small-step ongoing continual improvement activities to strategic breakthrough improvement projects. The laboratory should have a process in place for the identification and management of improvement activities.

> **8 MEASUREMENT, ANALYSIS AND IMPROVEMENT**
> **8.5 Improvement**
> 8.5.1 Continual improvement
> - Improvement of the effectiveness of the QMS through the use of the quality policy, quality objectives, audit results, analysis of data, corrective and preventive action and management review
> - Small-step improvements
> - Strategic breakthrough improvements
> - Process for the identification and management of improvement activities

Slide 62

Corrective action shall be used as a root for improvement. Sources of information for the consideration of corrective action include customer complaints, non-conformity reports, internal audit reports, output from management review, output from data analysis, outputs from satisfaction measurements, results of self-assessments etc. Corrective action usually entails significant cost, which is balanced against the impact of the problem being considered before the final decision to proceed with the appropriate corrective action.

The elements of the documented procedure are shown in this slide.

> **8 MEASUREMENT, ANALYSIS AND IMPROVEMENT**
> **8.5 Improvement**
> 8.5.2 Corrective action
> - Elimination of causes of non-conformity
> - Prevention of recurrence
> - Documented procedure
>   - Review of non-conformities and customer complaints
>   - Causes of non-conformities
>   - Need for action
>   - Records of the results of action taken
>   - Reviewing the effectiveness of the corrective action taken

Slide 63

Preventive action is taken to eliminate the causes of potential non-conformities in order to prevent their occurrence. Corrective action refers to the activities taken to correct something done (e.g. a non-conformity) whilst preventive action refers to the activities so that to avoid something not to be done. The elements of the documented preventive action procedure are shown in this slide.

8 MEASUREMENT, ANALYSIS AND IMPROVEMENT
8.5 Improvement

8.5.3 Preventive action
- Elimination of causes of potential non-conformity
- Prevention of occurrences
- Documented procedure
  - Determination of potential non-conformities and their causes
  - Evaluation of the need for action
  - Implementation of action
  - Records of the results of action taken
  - Reviewing the effectiveness of the preventive action taken

Slide 64

ISO 9000 as a quality management system sets out the basic requirements that are necessary for an organization to administer its management activities in an effective and efficient manner. Such requirements are classified in five basic categories as shown in this slide. The basic element of the quality management system is the identification of each management activity as a distinct and well defined process. All quality management system processes have to be identified and the laboratory should establish procedures on how to manage such processes. The quality management system procedures and work instructions should be systematically documented, together with evidence that all of the activities implemented under the quality system fulfilled their requirements.

Summary
- ISO 9000 basic requirements
  - Quality management system
  - Management responsibility
  - Resource management
  - Product realization
  - Measurement, analysis and improvement
- The process approach
- Documented quality management system
- Objective evidence of the conformance of the laboratory activities to the requirements

Slide 65

Where to Get More Information

- Quality Management Systems – Requirements
  ISO 9001:2008
- Quality Management Systems - Guidelines for performance improvements
  ISO 9004:2000
- http://www.iso.org

Bibliography

Arvanitoyiannis I, Hadjicostas E (2001) Quality Assurance and Safety Guide for the Food and Drink Industry; CIHEAM/Mediterranean Agronomic Institute of Chania / European Commission MEDA

Dux JP (1986) Handbook of Quality Assurance for the Analytical Chemical Laboratory; Van Nostrand Reinhold

Friedman FB (1995) Practice Guide to Environmental Management. 6th edition, Environmental Law Institute

ISO 9000:2005 Quality Management Systems – Fundamentals and Vocabulary

ISO 9001:2008 Quality Management Systems – Requirements

ISO 9004:2000 Quality Management Systems – Guidelines for Performance Improvements

Johnson PL(1993) ISO 9000 - Meeting the new international standard, McGraw Hill

Krajewski JL, Ritzman PL (1999) Operations Management, Strategy and Analysis. 5th edition, Addison-Wesley

Pierce FD (1995) Total Quality for safety and health professionals. Government Institutes USA

4 Accreditation or Certification for Laboratories?

Kyriacos C. Tsimillis

This presentation is focused on explaining the significance of accreditation and certification for laboratories and illustrates the usefulness of both procedures. The implementation of these procedures in laboratories is described, pointing out their similarities and differences. Reference is made to some publications. The discussion reflects the existing practice.

Slide 1

International agreements and regional cooperation e.g. GATT, European Acquis, specify the requirements for the free movement of goods and the elimination of technical barriers to trade, mutual recognition of test results and technical harmonization, in order to meet the needs of the market. The quality concept is being broadened to include additional aspects, tools and procedures.

Slide 2

This is reflected in both legislative requirements and greater competition to meet the demands of society. The access of consumers to more global and detailed information systems, referring to the particular features of products and services provided, as well as increasing consumer awareness, put additional pressure on manufacturers and suppliers.

B.W. Wenclawiak et al. (eds.), *Quality Assurance in Analytical Chemistry: Training and Teaching*, DOI 10.1007/978-3-642-13609-2_4, © Springer-Verlag Berlin Heidelberg 2010

Slide 3

Some specific needs refer to fields of particular importance, e.g. food quality including all relevant issues; consumer goods governed by the New Approach Directives and relevant harmonised European standards; the protection of the environment and the quality of life as well as the economic concerns of the consumers; dangerous substances and preparations and their impact on human beings, animals and the environment; forensic science.

Specific Needs...

exist in some sectors
- Food
- Health care services
- Environmental studies
- New Approach Directives
- Dangerous substances
- Forensic science

Slide 4

Conformity assessment can be considered as any action that is undertaken to determine whether a product, a service or a process meets specified requirements. It has been realised that the creation of an international technical language is of high priority; to this end emphasis has been given to enhance cooperation, both on regional and international levels in various fields related to conformity assessment activities.

There is a Need to...

create an international technical language to ensure common understanding!

Slide 5

Among them, standardization, testing, inspection and quality assurance activities have been developed to cover additional fields of products and services. At the same time great effort is being made worldwide to ensure a harmonised approach both in terminology and the context of various procedures.

As a Result...

efficient tools and mechanisms are required
- Emphasis is given on their worldwide implementation in a harmonised way

Slide 6

Specific tools and mechanisms are being used worldwide to meet the specified targets. They represent components of the international technical language referred to above (see slide 4). Depending on the particular application, some of these components e.g. standardization, quality and environmental management, accreditation, are being used on a voluntary basis while others are implemented on a compulsory basis within a relevant legislative framework; this is the case with Good Laboratory Practice (GLP) and New Approach Directives.

Tools and Mechanisms

- Standardization
- Laboratory infrastructure
- Quality management
- Metrology
- Accreditation
- GLP

Slide 7

New Approach Directives, as yet, cover up to 25 products or groups of products specifying the essential requirements with regard to health, safety and environmental issues. The concept of the New Approach in association with Technical Notification promotes aims related to the establishment of the single european market.

Tools and Mechanisms (cont.)

- Technical notification
- Environmental management
- Modules / notified bodies
- Harmonised standards
- Ecolabeling
- Occupational health and safety

Slide 8

All these tools and mechanisms require the appropriate infrastructure to be in place and efficiently operational. Specific tasks are undertaken by competent bodies and institutions in compliance with relevant standards and guidelines, which are issued by regional and international organizations.

The Infrastructure Required...

consists of
- Laboratories
- Standardization bodies
- Certification bodies
- Inspection bodies
- Accreditation bodies

Slide 9

How can suppliers and customers
worldwide be sure that products and
services meet relevant criteria? *When
they meet standard specifications.*
How can they be sure that products
and services meet such specifications?
By conformity assessment.
But how can they be sure that conformity assessment activities are reliable?
*When they comply with the criteria
laid down in ISO/IEC standards and
guides.*

How can We be Sure that...

products and services meet the
specifications?

- By "conformity assessment", i.e.
  checking that products and services
  meet the relevant standard's
  specifications

Slide 10

In official documents of the European
Commission it is stated that accreditation should be seen by public authorities, by the accreditors, by the certifiers
and by the industry as the highest level
of control of conformity assessment
activities, from a technical point of
view. The diagram illustrates the
correlation of various activities and the
role of the stakeholders. Accreditation
provides the highest level of recognition
of activities within conformity assessment.

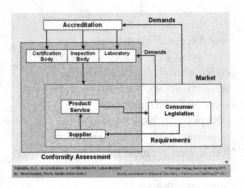

Slide 11

Definitions referring to these activities
are given in ISO/IEC Guide 2:2004,
EN 45020:2006 and some other normative documents, namely ISO/IEC
9000:2005 and ISO/IEC 17000:2004.
Definitions for accreditation and
certification given in ISO/IEC Guide 2
are used in the following discussion.

Definitions and Terminology

for standardization and related
activities
- are given in various standards
  - ISO/IEC 17000:2004
  - ISO/IEC Guide 2:2004
  - EN 45020:2006
  - ISO/IEC 9000:2005

Slide 12

Accreditation is the third-party attestation related to a conformity assessment body conveying formal demonstration of its competence to carry out specific conformity assessment tasks (ISO 17000).
Certification is the procedure by which a third party gives written assurance that a product, process or service conforms to specified requirements.

Following the Definitions...

- Accreditation refers to competence to carry out specific tasks
- Certification refers to assurance of conformity to specified requirements

Slide 13

Various activities come under conformity assessment; all of them are based on sampling, testing and reporting. First-party testing is carried out by manufacturers and suppliers. Second-party testing is performed by buyers, users or consumers. Third-party testing is carried out by organizations independent of the above parties.

Testing is Carried out...

by laboratories
- In the private sector
  - In-house
  - Second party
  - Third party
- In the public sector
  - Competent authorities

Slide 14

Several tens of thousands of standards have been published by national, regional and international organizations. All this accumulated know-how would remain of limited importance if no means were available for its use as the basis for measurements; measurement and testing are required to assess compliance! The laboratory infrastructure is based on the equipment, the methodology, the premises and the environmental conditions and the personnel involved.

We Need to Measure!

- If you cannot measure something, it doesn't exist!
- Therefore we need
  - Methodology
  - Equipment
  - Personnel

Slide 15

The role of laboratories is manifold. The main components are testing and calibration which require measurements and comparisons. This is the area to focus on in the discussion of "accreditation vs. certification". Other services provided, e.g. development of new methods, consultancy etc., may lead to different conclusions with regard to the relative importance of accreditation and certification.

The Role of Laboratories

- Drafting of standards
- Development of new methods
- Implementation of standards, technical regulations and the legislation
- Provision of services (testing, calibration, consultancy)

Slide 16

A laboratory requires a constant relationship and close co-operation with all other stakeholders. The laboratory is a living organism within a living and very competitive environment. The links to be established with all stakeholders should be clearly specified to enable the laboratory to function efficiently and meet the needs of its customers (the society through legislation and individuals). The laboratory is supplied with products (reagents, equipment, reference materials), receives services (accreditation, certification, provision of interlaboratory comparison schemes) and cooperates with other laboratories (cooperation in testing, provision of calibration services).

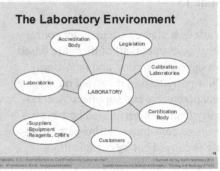

The Laboratory Environment

Slide 17

Why do the laboratories need to demonstrate their competence and the reliability of their results? Because they need to have competitive advantage and, at the same time, comply with legislation, where, more and more, reference is made to the need for reliable results and compliance with specific requirements (Accreditation, Good Laboratory Practice etc.).

Laboratories Need...

to demonstrate their technical competence and reliability

- To this end they also need to have policies how to estimate the uncertainty of their measurements

Slide 18

A series of factors are of decisive importance for the operation and efficiency of a laboratory. They are a mixture of both management and technical competence issues. Guidelines on how to deal with them are included in widely accepted documents.

Laboratories' Operation is Based on...

- Quality Management System
- Methodology
- Equipment (suitability, calibration, maintenance, conditions)
- Personnel (competence)
- Determination of uncertainty

Slide 19

A number of documents are very important for laboratories. Their utilisation may vary depending on the specific tasks of the laboratory; they are mainly standards/guides of wide acceptance e.g. from ISO, CEN, ILAC, European directives and EA guidelines, pieces of national and regional legislation etc. They are part of the laboratory's documentation system.

Basic Documents for Laboratories

- ISO/IEC 17025
- EA and ILAC Guidelines
- ISO 9000
- ISO 14000 (if applicable)
- GLP (if applicable)
- Legislation

Slide 20

Further to these, the laboratory community has established very useful links to exchange information and ideas, to submit suggestions and develop policies. Some of the most important fora on European and international level are EA (European cooperation for Accreditation), EURACHEM, EUROLAB, ILAC (International Laboratory Accreditation Cooperation).

To Avoid any Hidden Barriers...

to the free movement of goods, services and personnel, conformity assessment activities should be homogeneous and be implemented according to relevant Guides of ISO/IEC and EA/ILAC Guidelines

Slide 21

Relevant websites are very important sources of information. Those interested can have access to latest inputs and publications. Another very useful contact point is IRMM (the Institute of Reference Materials and Measurements) of the European Union. EUROMED Quality Programme (2004 – 2008) aimed to support Mediterranean countries to upgrade their infrastructure in the perspective of the establishment, by the year 2010, of a Free Trade Zone in the Euro-Mediterranean region.

Sources of Information

- www.european-accreditation.org
- www.eurachem.org
- www.eurolab.org
- www.ilac.org
- www.irmm.jrc.be
- www.euromedquality.org

Slide 22

Accreditation and Certification are components of the common technical infrastructure referred to above (see slides 8 and 10). They represent important mechanisms to facilitate free movement of goods. In a general sense they are dealing with similar issues, mainly because they assess and declare compliance with standards. However they present some major differences, due to the fact that they have different objectives. As a result the procedures to be followed in each case are also different.

Accreditation and Certification

present both
- Similarities related to the audit-based nature of the procedures and
- Differences with regard to the tasks and the resulting procedure

Slide 23

The main issue that should be taken into account is that certification and accreditation serve different tasks. The standards used in each case illustrate this difference. The ISO 9000 standards were originally developed to enable manufacturers to demonstrate to all interested parties that they are exercising control over the whole procedure they carry out and over its quality.

They are Both Related to…

quality assurance policies considered of high importance tools in today's society with regard to the free movement of goods

Slide 24

Furthermore it is well known that the ISO 9000 series of standards are applicable in all kinds of activity, both in manufacturing and service sectors; to this end, they can serve their purpose to demonstrate the company's quality management awareness and organization. This may apply also to laboratories; however it has nothing to do with the particular need of laboratories to demonstrate their technical competence with regard to specific tasks.

However...

this is not enough and does not fit the purpose in the case of laboratories!

Slide 25

The first documents dealing with the competence of laboratories were ISO/IEC Guide 25 and EN 45001 standard. They were issued to provide the necessary basis to be used by laboratories and by the bodies that assess their technical performance. In parallel, accreditation bodies undertook additional tasks with regard to the assessment of other activities (certification of products and systems) as specified in other EN 45000 series standards. The ISO/IEC 17025 has replaced EN 45001 (see slide 40), while other standards within EN 45000 series have been replaced by the ones in ISO/IEC 17000 series.

We Need Something More!

- We need tools for the assessment of the technical competence of laboratories to carry out specified tasks

Slide 26

Confusion does exist as a result of different needs as well as different levels of understanding the significance of the terms and the purpose of each procedure. The problem is how to avoid the confusion that exists among laboratories, customers and, more generally, the users of laboratory results.

Is there any Confusion...

on the use of these two procedures, their significance and their role within conformity assessment activities?
- If so, why does this happen?

Slide 27

What about purchasers, both individuals and the ones involved in public procurement, who still mention in invitations for tenders and contracts that the suppliers, including laboratories, should be certified against ISO 9001?

Can Both Activities…

be used by laboratories to document their competence and the reliability of their results?

- Do these two procedures represent equivalent criteria?

Slide 28

As a result of the differences in nature and objective of the two procedures, differences exist in the way the relevant assessments are carried out. These differences concern the planning and the content of the assessment, the competence of assessors/technical experts, the emphasis given to some basic issues and how the logo is used in each case.

Main Differences Refer to…

- The objective and content
- The documentation
- Assessors' competence
- Procedure of assessment
- The use of the logo
- How wide the scope is

Slide 29

Technical competence is dealt with only in accreditation and refers both to the methodology used and the personnel involved. In accreditation, validation of methods and expression of uncertainty are of major importance as well as the calibration of equipment and traceability of measurements.

Differences in the Procedure

Refer to the following issues
- Validation of test methods
- Expression of uncertainty
- Calibration of equipment
- Competence of personnel

Slide 30

What about the uncertainty of a measurement? Are clients of the laboratories aware of quality issues and how they can interpret uncertainties given in a test/calibration report? It seems that much more effort should be made to ensure an adequate understanding of such terms.

What about Uncertainty?

- How educated is the customer to consider properly and make correct use of the expressed uncertainty of measurement?

Slide 31

Reliability is a more "customer-friendly" term for uncertainty! Users of test results, including legislators and competent authorities do not always understand how uncertainty can be used to help them meet their needs. Laboratories find it difficult to communicate efficiently with them with regard to this. Based on the existing experience, it seems that not all customers of a laboratory are prepared to pay for a result that is "uncertain to some extent".

The Meaning of Uncertainty

is not well understood by the customers
- They may not choose a laboratory which produces results that are "uncertain" to some extent!

Slide 32

In accreditation the assessment team should consist of assessors/technical experts able to assess all relevant issues including the competence of the staff of the laboratory. This competence needs to be documented by relevant academic/technical qualifications, experience and training as well as successful participation in proficiency testing.

Technical Competence of Personnel Means...

- Suitable academic qualifications
- Adequate experience/training
- Awareness and familiarisation with particular methods and techniques

Slide 33

The basic question of whether accreditation or certification is appropriate arises when a laboratory is trying to meet the needs of the customers for testing and/or calibration services. Experience shows that the real nature and objectives of ISO 9001 are not always clearly understood by the stakeholders. The situation may be different in each case, depending on the customer.

Quality Means...

to meet the customers' needs
- How clear are these needs for the customers and how are they expressed in the communication with the laboratories?

Slide 34

Different groups of customers may introduce different requirements due to different levels of awareness of quality assurance and laboratory reliability issues. Competent authorities with a responsibility for the implementation of the legislation e.g. health, safety, environmental issues, organizations involved in procurements, the industry and individuals may have different requirements.

Who are the Customers?

- Competent authorities
- Industrial units
- Purchasers
- Individuals
- Other laboratories

Slide 35

One of the main reasons for the market demand for certified laboratories refers to the requirement of ISO 9001 for approved suppliers and subcontractors. Certified companies usually require that their potential subcontractors and suppliers are certified themselves. This may also apply for laboratories intended to offer their services to these companies.

ISO 9001 Provides...

for suppliers and subcontractors to be included in the approved list
- Relevant requirements are specified in the quality management system

Slide 36

Certification against ISO 9001 is the
usual requirement considered to meet
these needs. As a result the industrial
units and other enterprises are looking
for certified laboratories for testing
and calibration services. However, the
fulfilment of this requirement does not
demonstrate the competence of the
laboratory.

The Most Usual Requirement...

is ISO 9001 certification
* This refers also to laboratories that are
  listed as approved suppliers/
  subcontractors

Slide 37

A survey was carried out in 1996 by
ILAC to evaluate the status of laboratories worldwide with regard to
accreditation and certification as well
as the needs of their customers and the
trends in the market. Up to 500 laboratories responded leading to a series of
comments and suggestions.

A Survey was Carried out...

by ILAC in 1996 to evaluate the
competence of the laboratories, the
needs of their customers and market
trends

Slide 38

Only a small percentage (up to 10%)
hold an ISO 9001 certification (or
recognition). Big laboratories made
more use of their accreditation certificates to demonstrate compliance to
ISO 9001, but not very successfully
(up to 40%). Accreditation is considered
as more effective than certification to
consolidate market position and
demonstrate technical competence.

According to the Survey...

* A small percentage of laboratories were
  certified
* Big laboratories used accreditation for
  ISO 9001 compliance
* Accreditation was preferred to
  certification to demonstrate technical
  competence

Slide 39

The following are among the sugges-
tions made by a high percentage of the
laboratories:
-    Revision of the then existing tools
     (ISO/IEC Guide 25 and EN
     45001)
-    Harmonization of certification and
     accreditation documents with
     regard to management issues
-    Broadening of accreditation
     standard to cover management
issues for the whole activity of the laboratory
-    Encouraging cooperation and mutual recognition with regard to management
     issues.

Suggestions were then Made …

for the revision of ISO/IEC Guide 25 and
EN 45001 and harmonization with ISO
9001 on management issues

- This would facilitate the promotion of
  mutual recognition of this part of the
  system

Slide 40

The drafting of an ISO standard on
laboratory accreditation to be used
instead of EN 45001 was a need for
international trade. ISO/IEC Guide 25
could not be used for this purpose
since it was just a guide. However
other needs were also reflected in the
new standard, namely the ones related
to the question "accreditation vs.
certification for laboratories" and how to face problems arising from the existing
confusion.

The Standard ISO/IEC 17025:1999

- Has replaced the former European
  standard EN 45001 to ensure a
  worldwide acceptance of a unique
  document for the assessment of
  laboratories under accreditation

Slide 41

The fact that ISO 9001 is a generic
standard for quality management
applicable to all organizations, no
matter what the specific type of activity,
implies that a laboratory needs also to
meet quality management requirements in the way specified. The new
accreditation standard provides for this
in a very clear way, i.e. including relevant provisions.

1999 Edition of the Standard

- Provided for a clarification of the
  situation
- Classified the main requirements in two
  main categories
  - The management issues and
  - The technical competence issues

Slide 42

ISO/IEC 17025 classifies the main
requirements in two main categories,
namely the Management Issues (Chapter
4 in the standard) and the Technical
Competence Issues (Chapter 5 in the
standard). The new standard ISO/IEC
17025 was, in this respect, fully in line
with ISO 9001. A detailed correlation
with ISO 9001 and 9002 was included
in an Annex to the standard.

The Structure of the Standard...

enabled a direct correlation with ISO
9001:1994
- ISO/IEC 17025 specified management
  requirements (corresponding to the
  ones of ISO 9001) and technical
  requirements in separate chapters

Slide 43

The fact that the two standards can be
correlated and considered equivalent
with respect to the quality management requirements that they specify,
facilitates the mutual recognition of
systems based on them.

Therefore the Two Standards...

could easily be correlated and provide the
basis for comparison of the two
procedures they refer to

Slide 44

The main features and the points
mentioned during this presentation are
not affected by the revision of ISO in
2000. However this revision introduced the need for a revision of
ISO/IEC 17025 as well, soon after its
implementation, due to the fact that it
refers to the 1994 issue of ISO 9001.
Since the 1994 version was not in use
any more, there was a revision of
ISO/IEC 17025 in 2005.

The Standard ISO 9001:2000

Was the new version of the certification
standard
- As a consequence ISO/IEC 17025 was
  also revised to be again fully compatible
  with the new ISO 9001

Slide 45

This requires some cooperation between accreditation and certification bodies; such cooperation would be appreciated by laboratories since it would provide them with flexibility, lower costs and man-hours and reduce duplication in documentation and assessments.

It is Expected that...

in this way an agreement for mutual recognition could be established thus enabling laboratories to avoid duplication in work and additional costs

Slide 46

In order for a laboratory to achieve reliability and demonstrate its competence, it needs to proceed with some basic steps, starting with the preparation and implementation of a quality management system. Other steps are as follows…

To Achieve Reliability...

a laboratory needs to take some basic steps, starting with the preparation and implementation of a quality management system

Slide 47

The diagram illustrates the successive steps to be taken by a laboratory to establish reliability. The various steps refer to different levels of this procedure. First of all we need to ensure the organizational background to deal with all management issues. The second step includes issues related to the technical competence while the third step refers to the successful assessment that the target is met. The steps

correspond to different levels, e.g. the first step to certification while the third step to accreditation.

Slide 48

The basic starting point for a laboratory trying to establish reliability is the preparation and implementation of a quality management system. The standard to be used is either ISO 9001 where all management issues are addressed or ISO/IEC 17025. This step does not impose the need for certification and does not reflect the competence of the laboratory.

When Taking the First Step...

the Laboratory should prepare and implement a Quality Management System as described by either ISO 9001 or ISO/IEC 17025

- This step does not impose the need for certification

Slide 49

Proper documentation is required for all aspects dealt with so far in order to build confidence in the laboratory's results. Documentation is necessary to illustrate that every-day operation of the laboratory fully complies with the set requirements.

Further to a Quality Management System....

we need to ensure factors contributing to the reliability of the laboratory

- Technical competence of personnel
- Adequacy of infrastructure
- Fitness of methodology

Slide 50

A usual practice for laboratories is to include in the scope of their accreditation a selection of methods they use. The standard provides for certain rules to be followed in reports including results from other methods as well in any relevant declarations. However, the management issues addressed in the quality management system, following ISO/IEC 17025, cover the overall operation of the laboratory.

The Scope of Accreditation

- Usually includes only few of the methods carried out in the laboratory
- These methods are the ones for which technical competence is assessed
- The quality management system covers the overall activity

Slide 51

In the case of industrial units and other companies certified against ISO 9001 and operating their own laboratory, the latter, as part of the whole system, meets the relevant requirements of ISO 9001.

Laboratories Operating in...

certified industrial units have to meet the relevant requirements of ISO 9001
- Therefore they may find it easier to go on towards accreditation

Slide 52

The accreditation standard specifies that in the case of a laboratory operating as a part of a larger organization performing activities other than testing and/or calibration, appropriate arrangements should be made to avoid conflict of interest and ensure the independence and integrity of the laboratory.

However, in Such a Case...

one additional requirement should be considered, i.e. the one referring to the documentation of the independence, in practice, of the laboratory from the parent company

Slide 53

The wide acceptance of ISO 9001 and the confidence that specifiers and purchasers have in products and services provided by companies/organizations operating quality management systems according to ISO 9001 cannot be easily ignored. Laboratories have to meet the needs of their customers. Increase in awareness is still necessary.

It should be Clear that...

- If the need refers only to compliance with quality management issues, certification may be considered adequate
- Accreditation may also be considered as equivalent with respect to these issues

Slide 54

The main objective is to ensure that the needs are adequately described in each case and that all stakeholders (customers, regulators, competent authorities and other interested parties) have a clear understanding of their needs and how to meet them. The compliance with quality management issues can be met by certification as well as by accreditation (relevant provisions); the demonstration of technical competence is achieved only through accreditation.

It should also be Clear that...

if the need refers to the technical competence, accreditation of the laboratory with a detailed description of its scope is the appropriate means to demonstrate it

Slide 55

A development of the last years is expected to serve more efficiently the need for clarification. Following the joint ISO-ILAC-IAF Communiqué of June 2005, a statement could be included on accreditation certificates. According to this statement, the accredited laboratory has demonstrated both its technical competence for the defined scope and the operation of a quality management system.

Very Useful Development

- **Statement on accreditation certificate:** "This laboratory is accredited in accordance with the recognized International Standard ISO/IEC 17025:2005. This accreditation demonstrates technical competence for a defined scope and the operation of a laboratory quality management system". (refer joint ISO-ILAC-IAF Communiqué dated 18 June 2005)

Slide 56

Both the statement and the joint Communiqué were renewed in January 2009 to reflect the situation with the new version of ISO 9001:2008 (see following slides).

The Statement and Joint Communiqué

- Were renewed in January 2009 to reflect the revision of ISO 9001:2008

Slide 57

The consistency in technical competence is interrelated to the compliance with the management system as specified in ISO/IEC 17025:2005, thus illustrating that the standard itself aims at both aspects.

Joint ISO-ILAC-IAF Communiqué (1)

- "A laboratory's fulfillment of the requirements of ISO/IEC 17025:2005 means the laboratory meets both the technical competence requirements and *management system requirements* that are necessary for it to consistently deliver technically competent test results and calibrations...

(continued)

Slide 58

Furthermore, the equivalence in the way both the laboratory accreditation standard and the certification standard deal with the management system requirements illustrates that an accredited laboratory meets, in practice, these requirements as specified in ISO 9001. After a proposal made via Eurachem and EA to ILAC, a similar joint Communiqué was agreed and signed in September 2009 by ILAC, IAF and ISO for medical laboratories against ISO 15189:2007.

Joint ISO-ILAC-IAF Communiqué (2)

- ... The *management system requirements* in ISO/IEC 17025:2005 (Section 4) are written in language relevant to laboratory operations and meet the principles of ISO 9001:2000 *Quality management systems – Requirements* and are aligned with its pertinent requirements".

Slide 59

There is Still...

much more work to be carried out towards a better understanding and harmonization of the approaches with regard to laboratory work and the use of test and calibration results

Bibliography

EUROLAB Report No 3/98 (1998) "Accreditation and/or Certification?", Workshop in cooperation with EURACHEM, Delft

EUROLAB Technical Report 2/2002 "Customer satisfaction with European accreditation bodies" available from http://www.eurolab.org

IAF-ILAC-ISO Joint Communiqué on the Management Systems Requirements of ISO/IEC 17025:2005, January 2009, available from http://www.ilac.org

IAF-ILAC-ISO Joint Communiqué on the Management Systems Requirements of ISO 15189:2007, September 2009, available from http://www.ilac.org

ISO (1998) Development Manual 2: "Conformity assessment", 2nd edition

ISO (2000) "Aligning intentions and implementation in conformity assessment", ISO Bulletin 31:6

ISO 9001:2008, Quality management systems – Requirements

ISO/IEC 17000:2004, Conformity assessment – Vocabulary and general principles

ISO/IEC Guide 2:2004 / EN 45020:2006 - Standardization and related activities - General vocabulary

ISO/IEC 17025:2005 - General requirements for the competence of testing and calibration laboratories

ISO 9000:2005 - Quality management systems - Fundamentals and vocabulary

Tsimillis K, Michael S (2009) Suggestion to Eurachem and EA for a decision on ISO 15189 and ISO 9001, June 2009

5 Good Laboratory Practice

Evsevios Hadjicostas

Slide 1

The principles of Good Laboratory
Practice (GLP) in conjunction with the
principles of Total Quality Management (see chapter 6) ensure the quality
and reliability of the laboratory results,
which in turn help to ensure the protection of the environment and human
health and safety. A step further is the
accreditation of laboratories to ISO
17025 (see chapter 2) to perform
specified activities.

Good Laboratory Practice

- Good Laboratory Practice (GLP) is a
  quality system concerned with the
  organizational process and the
  conditions under which a study is
  planned, performed, monitored,
  recorded, archived and reported.

Compliance with the GLP principles covers many aspects of accreditation to ISO
17025. The European Union has adopted the procedures for Good Laboratory
Practice of the Organization for the Economic Cooperation and Development
(OECD) and calls on member states to take all measures necessary to ensure that
laboratories carrying out tests on chemical products comply with the GLP principles.
The OECD's GLP principles are part of the European Commission Directive
2004/10/EC (http://eur-lex.europa.eu).

B.W. Wenclawiak et al. (eds.), *Quality Assurance in Analytical Chemistry: Training
and Teaching*, DOI 10.1007/978-3-642-13609-2_5, © Springer-Verlag Berlin Heidelberg 2010

Slide 2

The GLP rules were first introduced in 1976 in the US by the Food and Drug Administration (FDA) for non-clinical laboratory studies after some irregularities in investigations and reports of pharmaceutical companies were revealed. The aim was to create a quality standard for laboratory investigations. Based on this all other industrialised countries followed. On the one side because this was necessary to export to US, on the other hand because it was recognised that trust in such investigations could be improved with such regulations.

History of GLP

- First in US after irregularities in pharmaceutical companies
- Other countries followed
  - To increase trust
  - To be able to export to US
- OECD
  - First decision in 1981
  - Binding guidelines in 1989

To reduce barriers to trade the Organisation for Economic Cooperation and Development worked on that topic and published its first decision in 1981. To introduce an internationally accepted system in all countries the OECD published "Guides for Compliance Monitoring Procedures for GLP" and "Guidance for the Conduct of Laboratory Inspections and Study Audits" in 1989 which are binding for all OECD member states.

Slide 3

The primary objective of laboratory personnel is to produce quality results that are mutually accepted. The harmonization of the laboratories to the GLP principles will benefit the global economy by the avoidance of the technical barriers to trade and the duplication of analytical determinations. In addition, GLP will improve the quality of life via the protection of human health and the environment.

Why GLP?

- Development of quality test data
- Mutual acceptance of data
- Avoid duplication of data
- Avoid technical barriers to trade
- Protection of human health and the environment

Slide 4

The principles of GLP ensure the
generation of high quality and reliable
test data related to the safety of industrial chemicals, pesticides, pharm-
aceuticals, food and feed additives,
cosmetics, veterinary drugs as well as
food additives, in the framework of
harmonizing testing procedures for the
mutual acceptance of data.

Scope

- Non-clinical safety testing of test items
  contained in
  - Pharmaceutical products
  - Pesticide products
  - Cosmetic products
  - Veterinary drugs
  - Food and feed additives
  - Industrial chemicals

The purpose of testing these test items is to obtain data on their properties and/or
their safety with respect to human health and/or the environment.

Slide 5

In Europe *Directive 2004/9/EC*, on the
inspection and verification of good
laboratory practice, lays down the
obligation of the Member States to
designate the authorities responsible
for GLP inspections in their territory.
It also comprises reporting and internal
market (= mutual acceptance of data)
requirements. The directive requires

**The European Commission
Directive 2004/9/EC**

- EU Member States shall verify the
  compliance with GLP of any testing
  laboratory within their territory claiming
  to use GLP in the conduct of tests on
  chemicals

that the OECD Revised Guides for Compliance Monitoring Procedures for GLP
and the OECD Guidance for the Conduct of Test Facility Inspections and Study
Audits must be followed during laboratory inspections and study audits.

Slide 6

The Annex I of the Directive
2004/9/EC consists of two parts; Part
A (Guides for compliance monitoring
procedures for good laboratory practice) and Part B (Guidance for the
conduct of test facility inspections and
study audits). The provisions for the
inspection and verification of GLP

Inspection and Audit

- The authorities of the EU Member
  States shall inspect the laboratory and
  audit the studies in accordance with the
  provisions laid down in Annex I of the
  Directive 2004/9/EC

which are contained in Parts A and B are those contained in Annexes I and II
respectively of the OECD Council Decision-Recommendation on compliance with
principles of good laboratory practice.

Slide 7

To facilitate the mutual acceptance of test data generated for submission to Regulatory Authorities of the OECD member countries, harmonisation of the procedures adopted to monitor GLP compliance, as well as comparability of their quality and rigour, are essential. The aim of this part of the Annex is to provide detailed practical guidance to the Member States on the structure, mechanisms and procedures they should adopt when establishing national GLP compliance monitoring programmes so that these programmes may be internationally acceptable.

**Directive 2004/9/EC
Annex I, Part A**

- Guides for compliance monitoring procedures for good laboratory practice

Slide 8

Part B of Annex I, provides guidance for the conduct of test facility inspections and study audits which would be mutually acceptable to OECD member countries. It is principally concerned with test facility inspections, an activity which occupies much of the time of GLP inspectors. Test facility inspections are conducted to determine the degree of conformity of test facilities and studies with GLP principles and to determine the integrity of data to assure that resulting data are of adequate quality for assessment and decision-making by national Regulatory Authorities. They result in reports which describe the degree of adherence of a test facility to the GLP principles. Test facility inspections should be conducted on a regular, routine basis to establish and maintain records of the GLP compliance status of test facilities.

**Directive 2004/9/EC
Annex I, Part B**

- Guidance for the conduct of test facility inspections and study audits

Slide 9

OECD Member States developed the *OECD principles of GLP*, utilising common managerial and scientific practices and experience from various national and international sources. The purpose of these principles of good laboratory practice is to promote the development of quality test data. Comparable quality of test data forms the basis for the mutual acceptance of data among countries. If individual countries can confidently rely on test data developed in other countries, duplicative testing can be avoided, thereby saving time and resources. The application of these principles should help to avoid the creation of technical barriers to trade, and further improve the protection of human health and the environment. GLP principles are explicitly presented below.

The Commission Directive 2004/10/EC

- The harmonisation of laws, regulations and administrative provisions relating to the application of the principles of good laboratory practice and the verification of their applications for tests on chemical substances

Slide 10

Good Laboratory Practice is a quality system concerned with the organisation of the test facility and the conditions under which non-clinical health and environmental safety studies are planned, performed, monitored, recorded, archived and reported. This is the GLP Decalogue stating the basic principles that the laboratory must follow.

The GLP Principles

1. Test facility organization and personnel
2. Quality Assurance (QA) program
3. Facilities
4. Apparatus, materials and reagents
5. Test systems
6. Test and reference items
7. Standard Operating Procedures (SOP's)
8. Performance of the study
9. Reporting of study results
10. Storage and retention of records and materials

Slide 11

A broader term than "the laboratory" is the "test facility". This includes the persons, premises and operational units that are necessary for the operation of the laboratory. Test sites may be at different locations at which different phases of a study are conducted. Sponsor is the laboratory customer.

Terms Concerning the Organization of a Test Facility

- Test facility
- Test site
- Test facility management
- Test site management
- Sponsor
- Study director
- Principal investigator
- Quality assurance program
- Standard operating procedures
- Master schedule

Study director means the individual responsible for the overall conduct of the study. Principle investigator means the individual who acts on behalf of the study director in case of multi site studies. Master schedule means a compilation of information to assist in the assessment of the workload and the tracking of studies at a test facility.

Slide 12

Study means an experiment or set of experiments in which a test item is examined. The study plan defines the objectives and experimental design of the study.
Test system is the system under study, that is, what we are testing. Usually, we test or analyse a specimen i.e. a sample taken from the test system.

Terms Concerning the Study

- Non-clinical health and environmental safety study
- Short term study
- Study plan
- Study plan amendment
- Study plan deviation
- Test system
- Raw data
- Specimen
- Experimental starting date
- Experimental completion date
- Study initiation date
- Study completion date

Slide 13

Test item is the article that is the subject of a study. The reference item provides the basis for comparison. Batch means a specific quantity of the test or reference item expected to have a uniform character. Vehicle is the carrier of the test item.

Terms Concerning the Test Item

- Test item: the article that is subject of a study
- Reference item
- Batch
- Vehicle

Slide 14

The first principle of the GLP Decalogue is the organization and the personnel of the Test Facility, i.e. the entire laboratory. The responsibilities of the top management and those of the directors within the whole hierarchy are mentioned in detail below.

1. Test Facility Organization and Personnel

- Test facility management's responsibilities
- Study director's responsibilities
- Principal investigator's responsibilities
- Study personnel's responsibilities

Slide 15

The laboratory management has the responsibility to ensure that:
The principles of Good Laboratory Practice are complied within its laboratory, a *Laboratory Policy* exists and it is communicated and understood by the laboratory personnel, a sufficient number of *qualified personnel*, appropriate *facilities*, *equipment*, and *materials* are available in the laboratory and proper training and relevant

1. Test Facility Organization and Personnel
Test Facility Management's Responsibilities

• The management should ensure that
  • The principles of GLP are complied with
  • A sufficient number of qualified personnel, appropriate facilities, equipment and materials are available
• Records of qualifications, job descriptions, training and experience of personnel are maintained
• Personnel understand the functions they are to perform

records of qualifications, training and experience, as well as job descriptions for each professional and technical individual are maintained,
all *personnel* involved in the conduct of the laboratory activity must be knowledgeable about those parts of the principals of Good Laboratory Practice, which are applicable to the activity that they perform.

Slide 16

Technically valid *Standard Operating Procedures* (SOPs) have to be established and followed. In case of deviations, the impact to the quality is assessed and appropriate corrective and preventive action is taken. All SOP documents have to be controlled and a historical file has to be maintained. The management has to ensure that a *Quality Assurance Program* is in place and designated personnel perform the

1. Test Facility Organization and Personnel
Test facility Management's Responsibilities

• Appropriate and valid SOP's are established and followed
• A Quality Assurance Program is in place
• A Study Director and a Principal Investigator, if needed, is designated
• Documented approval of the study plan
• The study plan is available to quality assurance personnel

program in accordance with the principals of GLP.
A documented, properly maintained and approved (by dated signature) *Plan of the Laboratory Activities*, has to be in place and made available to laboratory and quality assurance personnel. The director of the laboratory is responsible for the overall conduct of the laboratory activities and the approval of the final reports.

Slide 17

Appropriate characterization has to be
made of reference substances, reagents
and other purchased items as well as
of test samples, solutions and other
preparations that are used for the laboratory activity. The supplies must meet
specified requirements appropriate to
their use.
Clear lines of communication have to
exist across the whole hierarchy and
between the director, laboratory personnel and the quality assurance team.

1. Test Facility Organization and Personnel
Test Facility Management's Responsibilities

- A document control system is in place
- Purchased materials meet specified requirements
- Test and reference items are appropriately characterized
- Clear lines of communication exist
- Computerized systems are suitable for their intended purpose

Computerized systems must be suitable for their intended purpose and be
validated. All data, including electronic data, must be controlled and properly
archived.

Slide 18

The Study Director is the single point
of the study control and has the
responsibility for the overall performance of the study and the final report.
He or she is the delegated individual
from the laboratory management to
make sure that the laboratory performs
under the GLP principles. His or her
responsibilities are shown on this and
the next slide.

1. Test Facility Organization and Personnel
Study Director's Responsibilities

- Has the responsibility for the overall performance of the study and the final report
- Approves the study plan and amendments and communicate them to the QA personnel
- Ensures that SOPs, study plans and their amendments are available to study personnel
- Ensures that the SOPs are followed, assess the impact of any deviations and takes appropriate corrective and preventive action

Slide 19

Study director's responsibilities
(continued)

1. Test Facility Organization and Personnel
Study Director's Responsibilities

- Ensures that
  - Raw data are documented and recorded
  - Computerized systems are validated
  - SOPs are followed
  - Deviations are acknowledged
  - Records and data are archived
- Sign and date the final report to indicate acceptance of responsibility

Slide 20

The laboratory personnel have their own responsibilities, i.e. to comply with the instructions given in the relevant documentation, for recording raw data promptly and accurately and are responsible for the quality of their results. The laboratory personnel have the responsibility to exercise safety precautions, to minimize risk to themselves and to ensure the integrity of their results. They must be aware of the risk associated with the laboratory work and have knowledge of the appropriate actions to be taken in order to protect all personnel having access to the laboratory (including subcontractors and visitors).

1. Test Facility Organization and Personnel
Study Personnel Responsibilities

- Knowledge of the GLP principals
- Access to the study plan and appropriate SOPs
- Comply with the instructions of the SOPs
- Record raw data
- Study personnel are responsible for the quality of their data
- Exercise health precautions to minimize risk
- Ensure the integrity of the study

Slide 21

The laboratory should have a documented *Quality Assurance Programme*, carried out by designated individual(s), to assure that the activities performed are in compliance with the principles of GLP. The QA personnel must conduct inspections to determine if all of the laboratory activities are conducted in accordance with the principals of GLP and defined SOPs. These personnel should not be involved in the laboratory activity being inspected.

2. Quality Assurance Program
General

- Documented Quality Assurance (QA) Program
- Designated individuals as members of the QA team directly responsible to the management
- QA members not to be involved in the conduct of the study being assured

Slide 22

The role of the quality assurance personnel is to inspect the laboratory activity and verify that this activity complies with the GLP principles. They have access to the study plans and Standard Operating Procedures and all updated versions. QA personnel verify in a documented way the compliance of the study plan with the GLP principles.

2. Quality Assurance Program
Responsibilities of the QA Personnel

- Access to the updated study plans and SOPs
- Documented verification of the compliance of study plan to the GLP principals
- Inspections to determine compliance of the study with GLP principles. Three types of inspection
  - Study-based inspections
  - Facility-based inspections
  - Process-based inspections

The inspection is performed according to the quality assurance programme and is based on the audit plan. There can be three types of audit:

Test- or Study-based, that is, to inspect thoroughly and meticulously a specific test, test method or study,

Facility-based, that is, to inspect the whole facility whether it meets the GLP requirements,

Process-based, that is, to inspect the process that the laboratory uses to perform a specific activity.

The inspection results are promptly reported in writing and addressed to the top-level management.

Slide 23

As part of the inspection report the QA personnel specifies the inspection types and inspected phases and confirms that the final report reflects the raw data.

The inspection results are reported to the management.

2. Quality Assurance Program
Responsibilities of the QA Personnel

- Inspection of the final reports for accurate and full description
- Report the inspection results to the management
- Statement

Slide 24

The laboratory area should be of suitable size, construction and location to meet the requirements of the laboratory activity and to minimise any disturbance that would interfere with the validity of the results.

3. Facilities

- Suitable size, construction and location
- Adequate degree of separation of the different activities
- Isolation of test systems and individual projects to protect from biological hazards
- Suitable rooms for the diagnosis, treatment and control of diseases
- Storage rooms for supplies and equipment

Slide 25

An adequate degree of separation of the different activities should be provided, i.e. facilities for handling the test samples and reference substances. Archive facilities should be provided for the secure storage and retrieval of raw data, final reports etc.
Samples, reagents, measurement standards and reference materials must be stored so as to ensure their integrity.
Handling and storage of wastes should be carried out in such a way as not to jeopardise the integrity of the laboratory activities.

3. Facilities

- Separate areas for receipts and storage of the test and reference items
- Separation of test items from test systems
- Archive facilities for easy retrieval of study plans, raw data, final reports, samples of test items and specimen
- Handling and disposal of waste in such a way not to jeopardize the integrity of the study

Slide 26

Laboratory apparatus should be periodically inspected, cleaned, maintained, and calibrated according to standard operating procedures.
Records of these activities should be maintained. Calibration should be traceable to national or international standards of measurements.
Chemicals, Reagents, and Solutions should be labelled to indicate identity (with concentration, if appropriate),

4. Apparatus, Materials and Reagents

- Apparatus of appropriate design and adequate capacity
- Documented inspection, cleaning, maintenance and calibration of apparatus. Calibration to be traceable to national or international standards
- Apparatus and materials not to interfere with the test systems
- Chemicals, reagent and solutions should be labeled to indicate identity, expiry and specific storage instructions.

expiry date and specific storage instructions. Information concerning source, preparation date, the person responsible for the preparation and the stability should be available. The expiry date may be extended on the basis of documented evaluation or analysis.
The correct disposal of chemicals, reagents and solvents is a matter of good laboratory practice and should comply with national and/or environmental and/or health & safety regulations.

Slide 27

Test systems could be physical, chemical or biological. When dealing with biological test systems e.g. animals or plants, special care should be paid to ensure proper conditions for storage, housing and handling. The GLP provisions regarding storage, housing, handling and care of biological test systems are shown in this and the next slide.

5. Test Systems

- Physical and chemical test systems
  - Appropriate design and adequate capacity of apparatus used for the generation of data
  - Integrity of physical/chemical test systems
- Biological test systems
  - Proper conditions for storage, housing, handling and care
  - Isolation of newly received animal and plant test systems until health status is evaluated
  - Humanely destruction of inappropriate test systems

Slide 28

5. Test Systems

- Records of source, date of arrival, and arrival conditions of test systems
- Acclimatization of biological systems to the test environment
- Proper identification of test systems in their housing or container or when removed
- Cleaning and sanitization of housings or containers
- Pest control agents to be documented
- Avoid interference from past usage of pesticides

Slide 29

Receipt, handling, sampling and storage of test and reference items should be clearly defined activities, with well-defined procedures. Also procedures must be provided for the records that are to be kept as evidence of good practice and the identification information on storage containers.

6. Test and Reference Items

- Receipt, handling, sampling and storage
  - Records for date of receipt, expiry date, quantities received and used in studies etc.
  - Handling, sampling and storage procedures to ensure homogeneity and stability and avoid contamination or mix-up
  - Identification information on storage containers

Slide 30

Characterization of test and reference items.

6. Test and Reference Items
- Characterization
  - Identification of each test and reference item
    - Code, CAS number, name etc.
  - Identification of each batch of the test or reference items
    - Batch number, purity, composition, concentration etc.
  - Cooperation between the sponsor and the test facility
    - Verification of identity of the test item

Slide 31

Characterization of test and reference items

6. Test and Reference Items

- Known stability of test and reference items
- Stability of the test item in its vehicle (container)
- Experiments to determine stability in tank mixers used in the field studies
- Samples for analytical purposes for each batch

Slide 32

The laboratory should have written Standard Operating Procedures approved by the Laboratory Director that are intended to ensure the quality and integrity of the data generated by that laboratory. The Laboratory Director has the responsibility to approve revisions to standard operating procedures. The SOPs must be available to the personnel that are involved in the relevant activities as well as to the quality assurance team that performs the audit.

7. Standard Operating Procedures

- Approved SOPs to ensure the quality and integrity of the laboratory data
- Immediately available current SOPs relevant to the activities being performed
- Deviations from SOPs to be acknowledged by the study director

Slide 33

Standard operating procedures should be available for, but not be limited to, the categories of laboratory activities shown in this and the next slide.

7. Standard Operating Procedures
- SOPs for
  - Test and reference items
    - Receipt, identification, labeling, handling, sampling, storage
  - Apparatus
    - Use, maintenance, cleaning, calibration
  - Computerized systems
    - Validation, operation, maintenance, security, change control, back-up
  - Materials, reagents and solutions
    - Preparation and labeling

Slide 34

7. Standard Operating Procedures
- Record keeping, reporting, storage and retrieval
  - Coding system, data collection, preparation of reports, indexing system, handling of data
- Test system
  - Room preparation, environmental room conditions, receipt, transfer, identification etc., test system preparation, observations etc.,
- Quality Assurance Procedures
  - Operation of QA personnel

Slide 35

The laboratory activities are performed following a well-defined laboratory plan that gives guidelines to personnel regarding
- the methodology to follow,
- the expected control limits of the parameters under study,
- the reference substances, reagents and solutions to be used,
- the instrumentation to be employed with reference to their performance parameters.

8. Performance of the Study
- Study plan
  - Written plan, verified for GLP compliance, approved by the study director and by the management
  - Approval of amendments by dated signatures
  - Deviations to be explained and acknowledged

Amendments of the study plan are approved and deviations from this are described, explained and acknowledged.

Slide 36

Content of the study plan

8. Performance of the Study

- Content of the study plan
  - Identification of the study
    - Title, nature and purpose of the study, test item identity, reference item used etc.
  - Information concerning the sponsor and facility
    - Names and address (sponsor, test facility, study director)
  - Dates
    - Approval dates of the study plan, estimated starting and completion dates etc.
  - Reference to test methods
  - Records

Slide 37

Each study is given a unique identification and it is conducted in accordance with the study plan. The analyst that performs the tests has the responsibility to record the laboratory data promptly, accurately and legibly. Any changes in the raw data should be made so as not to obscure the previous entry, should indicate the reason for change and should be dated and signed or initialled by the person making the change.

8. Performance of the Study

- Conduct of the study
  - Identification of each study
  - The study to be conducted in accordance with the study plan
  - Data generated to be recorded directly and accurately
  - Changes in the raw data not to obscure the previous data
  - Identification of electronic data

Slide 38

A final report should be prepared after the completion of each laboratory activity. The final report should be signed and dated by the laboratory director to indicate acceptance of responsibility for the validity of the data. The extent of compliance with the principles of good laboratory practice should be indicated. Corrections and additions to a final report should be in the form of amendments and a clear statement of the reasons for the corrections or additions should be made. The final report should include, but not be limited to, the information shown in this and the next slide.

9. Reporting of Study Results

- General
  - Final report for each study
  - Scientists to sign and date their reports
  - Approval by the Study Director
  - Corrections, additions, amendments to be signed and dated by the study director
- Content of the final report
  - Identification of the study
    - Descriptive title, identification of the test and reference item, purity, stability

Slide 39

Content of the final report

Slide 40

A system is to be developed to define the way that records and materials are stored and retained. Records that are stored include, but are not limited to, those on this slide.

Slide 41

In the absence of a required retention period, the final disposition of materials is documented. When samples are disposed of before expiration date for any reason, this should be justified and documented.

Material retained in archives should be indexed so as to facilitate orderly storage and retrieval. Only authorized personnel have access to the archives.

Slide 42

Good Laboratory Practice is a quality
system that is tailored to the needs of
the modern laboratories. Its principles
are very close to the principles of the
modern quality management systems
like ISO 9000 and ISO 17025. However, GLP is specific to the non-
clinical health and environmental
safety studies, it covers physical and
chemical test systems, and gives emphasis to biological test systems. It is
of utmost importance to note that GLP principles were set out by the Organization
for Economic Cooperation and Development (OECD) and after being modified
they were adopted by the European Union and are now the Commission Directive
1999/11/EC.

Summary

- GLP v/v ISO 9000 and ISO 17025
- Non-clinical health and environmental safety studies
- Physical and chemical test systems
- Biological test systems
- OECD
- EU Directive 1999/11/EC

Slide 43

Where to Get More Information
- Commission Directive 2004/9/EC and Commission Directive 2004/10/EC
  - http://eur-lex.europa.eu
- http://ec.europa.eu/enterprise/chemicals /legislation/glp/index_en.htm
- http://www.oecd.org under environment/chemical safety

Bibliography:

Anderson MA (2000) GLP Quality Audit Manual. Informa Healthcare

Anderson MA (2002) GLP Essentials – A Concise Guide to Good Laboratory
Practice. Informa Healthcare

Arvanitoyiannis I, Hadjicostas E (2001) Quality Assurance and Safety Guide for
the Food and Drink Industry. CIHEAM/Mediterranean Agronomic Institute of
Chania/European Commission MEDA

Crosby NT, Patel I (1995) General Principles of Good Sampling Practice. The
Royal Society of Chemistry

Dux JP (1986) Handbook of Quality Assurance for the Analytical Chemical Laboratory. Van Nostrand Reinhold

Garner WY, Barge MS (1989) Good Laboratory Practices, An Agrochemical Perspective. ACS Symposium Series 369

Haider SI (2002) Pharmaceutical Master Validation Plan: The Ultimate Guide to FDA, GMP, and GLP Compliance. Informa Healthcare

Seiler JP (2005) Good Laboratory Practice: The Why and the How, Springer, Berlin

6 Total Quality Management and Cost of Quality

Evsevios Hadjicostas

Before we start analysing the philosophy of Total Quality Management it is worthwhile going back to the early days of quality and the quality movement. In fact, the quality concept dates back to the creation of Adam and Eve: "*And God saw every thing that he had made, and, behold, it was very good*". (Genesis A 31). It is remarkable that at the end of each day, looking at his creations God was saying, "*This is good*". However, at the end of the sixth day, after he finished the creation of human beings, he said, "*This is very good*". It is amazing that he did not say, "*This is excellent*". This is because excellence is something that we gain after tireless effort. God left room for improvement in order to challenge us and make our life more attractive, which has really happened!

In the text that follows we will be referring to the term "*organization*" and we mean the laboratory. The term "*process*" refers to any activity in the organization, e.g. the analytical process.

Slide 1

Before mass production the control of quality was in the hands of the operator. This period, of the "*operator's Quality Control*" was followed by the "*foreman's Quality Control*" period, where a supervisor was assigned to supervise the quality of the work of the operators. Then it was the "*Full-time inspectors Quality Control*" (1920-1930) and the

The International Quality movement

- Operator Quality Control
- Foreman Quality Control
- Full-time Inspectors
- Statistical Quality Control
- Total Quality Control
- Total Quality Management

fourth stage was the "*Statistical Quality Control*" (SQC), first introduced by Deming in Japan in early 1950s.

During 1950 Feigenbaum introduced the concept of *Total Quality Control* (TQC) into western countries, as a management tool to improve product design and quality by reducing operating costs and losses. Kauro Ishikawa, a Japanese chemist, introduced TQC in Japan around 1950-1960 as "*a company-wide management tool in order to produce high quality goods and services that give a competitive edge both in the short term and in the future*".

B.W. Wenclawiak et al. (eds.), *Quality Assurance in Analytical Chemistry: Training and Teaching*, DOI 10.1007/978-3-642-13609-2_6, © Springer-Verlag Berlin Heidelberg 2010

Slide 2

Total Quality Management (TQM) is a company culture that allows it to provide quality goods and services at the lowest cost in order to achieve customers' satisfaction and, at the same time, ensure satisfactory business development by continuous improvements. The definition of TQM shown in this slide is taken from the British Standard 7850.

Total Quality Management

Total Quality Management (TQM) is a philosophy and involves company practices that aim to harness the human and material resources of an organization in the most effective way to achieve the objectives of the organization   (BS 7850)

In this chapter we present the management principles of Total Quality Management and the quality improvement methods, as set out in the standards BS 7850 part 1 and part 2 respectively. Furthermore, we introduce the cost of quality and give an outline of the economics of quality, as set out in the BS 6143, Guide to the economics of quality.

Slide 3

For an organization to function effectively and efficiently, it has to identify numerous linked activities. An activity using resources, and managed in order to enable the transformation of inputs into outputs, is considered as a process. Often the output from one process directly forms the input to the next.

What is a Process?

Any activity that accepts inputs, adds values to these inputs for customers, and produces outputs for these customers. The customer may be either internal or external to the organization

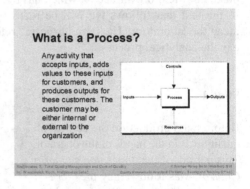

Slide 4

The process of TQM consists of three major activities:
- Setting the policy and strategy of the organization,
- effectively and efficiently managing the organization, and
- seeking continuous improvement.

Typical Process of TQM

- **Policy and strategy of the organization**
  - Mission
  - Leadership and commitment
  - Divisional objectives
- **Management of the organization**
  - Organization structure
  - Management system
  - Information system
  - Communication
- **Improvement of the organization**
  - Working environment
  - Measurement of performance
  - Improvement objectives
  - Improvement plans
  - Monitor and review

Slide 5

Firstly, the fundamental concepts of Total Quality Management are introduced. When this is understood the question that arises is *"what can we do to implement the TQM principles in our organization?"*. At this point we will understand the need to explain further the meaning and fundamentals of the *"quality improvement"* and how to manage quality improvement. A set of tools for quality improvement has to

Presentation Outline

- Fundamental Concepts
- Implementing Total Quality Management
- Quality Improvement
- Managing for Quality Improvement
- Tools for Quality Improvement
- Quality Gurus
- Cost of Quality

be installed in our organization so that it is adequately equipped for quality improvement and TQM. To understand TQM in practice a few quality philosophers and their approach to quality will be presented. The ultimate goal of TQM is to give benefits to the organization e.g. by minimizing the cost of quality and maximizing productivity.

Slide 6

The philosophy of the Total Quality Management is applicable to any organization, which is customer oriented and committed to quality. Management commitment to Total Quality is essential for the organization to achieve excellence.
The commitment to quality must be conveyed to all levels and activities of the organization.
Furthermore, management commit-

Fundamental Concepts
Commitment to TQM

- Commitment to TQM by the highest level of management
- Promotion of this concept to all levels and activities of the organization
- Individual involvement
- Devotion to continuous improvement

ment involves every department, function and process in the organization and the active commitment of everyone in the organization to meeting customer needs and seeking continuous improvement.

Slide 7

All individuals in the organization must be aware of the customers' needs and be actively committed to satisfying customers' requirements. The laboratory's customer is the *raison d' être* of the laboratory and he/she must be satisfied and be proud of his/her supplier.

An internal customer is a member of the organization, which is part of the process and undertakes the performance of a specific task. The *"supplier"* to this internal customer is again a member of the organization that transfers his/her product or service to the next member of the chain (his/her customer). To do things *"right first time"* and on time, and achieve (external) customer satisfaction the internal suppliers and internal customers must always work in perfect coordination.

Fundamental Concepts
Customer Satisfaction

- Internal customers
- External customers

- Customer needs
- Customer expectations

Slide 8

The resources that are supplied to the laboratory process must be utilized in a way that maximizes the benefit to the laboratory and minimizes the loss. Quality losses are caused by the failure to utilize most effectively the potential of human, financial and material resources in a process. This could be

- due to loss of customer satisfaction,
- due to loss of the opportunity to add more value to the customer, the organization or the society and/or
- due to waste or misuse of resources (e.g. peoples' health, damage to property, process interruption).

Opportunities to reduce quality losses steer quality improvement activities.

Fundamental Concepts
Quality Losses

- Ineffective and inefficient utilization of human, financial and material resources in processes
- Loss of customer satisfaction
- Loss of opportunity to add more value
- Loss due to waste or misuse of resources

Slide 9

Quality is everyone's job. The laboratory personnel must have a construc-
tive role in all laboratory functions and
processes. They must operate with a
team spirit and enhance the synergistic
effect of teamwork. All laboratory
members are links in the chain and
nodes in the quality grid. Their
strengths and abilities must be fully
and effectively utilized. Within this
concept the laboratory management
recognizes and makes use of the strengths of the laboratory personnel and at the
same time recognizes and improves their weak points. It is the responsibility of the
chief executive and the laboratory management to cultivate team spirit and make
the merits of the human resource known to the laboratory.
Communication and teamwork remove organizational and technical barriers and it
is essential for everyone to identify and follow up opportunities for improvement.
They should be extended through the whole supply chain including suppliers and
customers.

Fundamental Concepts
Participation by All

- Strengths and abilities of individuals
- Effective utilization of strengths and abilities
- Communication and teamwork

Slide 10

Measurement is the proof of improvement. Quality improvement cannot be
verified if quality is not measured. The
very first step for a quality-oriented
laboratory is to establish its present
situation and quality level. Future
comparisons will be referred to this
base level. Then, the laboratory must
define the parameters that the measurement is based on. Quality is monitored
on a systematic basis and the quality
results are recorded.

Fundamental Concepts
Process Measurements

Slide 11

Quality improvement is the aim of any modern organization. Improvement needs to be continuous and never ending. The means of improvement can be applied to both the staff and the processes. The degree of improvement achieved can be measured, monitored and controlled.

Slide 12

Problems are present in all real situations. These problems provide opportunities for improvement. Problems are there to give the laboratory the opportunity to solve them and to enable laboratories to learn from their problem solving experiences. Provisions for identification and resolution of potential and existing problems on a continuous basis are essential.

Slide 13

The corporate objectives of a laboratory are conveyed to the laboratory members via a well-defined human resource appraisal and development system. The appraisal process is beneficial for both the laboratory management and the laboratorystaff. The goals assigned to the individuals are

-    customer oriented and
-    in line with the corporate goals and the mission of the laboratory.

Fundamental Concepts
Alignment of Corporate Objectives and Individual Attitudes

- Appraisal and development of human resource
- Quality improvement goals
  - Clear, understandable, specific, measurable, achievable, realistic, time-bound and agreed to by all relevant individuals.
  - Focused on customer satisfaction
  - In line with overall business goals and mission

Slide 14

All laboratory members should accept and recognize the individual's responsibility and authority. This responsibility and authority of each individual is defined in job description documents, which are agreed with the interested parties and communicated to all relevant laboratory sectors.

Fundamental Concepts
Personal Accountability
Personal Development

- Responsibilities and authorities for all individuals
- Job descriptions
- Appraisal, training and development of individuals at all levels

Laboratory management must provide for the continuous appraisal, training and development of individuals at all levels in the organization. Training and education is, in fact, part of the remuneration package for all level of employees in the laboratory, as well as for self-employed chemists.

Slide 15

Now that we understand the fundamental concepts of TQM we proceed to the prerequisites needed to implement TQM.
The laboratory should develop and apply a range of appropriate systems, improvement tools and techniques and coordinate their operation in order to utilize their full potential and achieve improvement.

Implementing Total Quality Management

- Appropriate systems, improvement tools and techniques
- Application and coordination of the above
- Overcome resistance to change

Very often, there is a resistance to change, which requires particular attention to be overcome. It is the responsibility of the management to make people understand that any improvement needs changes to the existing practices and that the benefits from improvement will satisfy all the organization's members.

Slide 16

Resistance to change is overcome and quality management is achieved through the ongoing incremental improvement of the laboratory processes.

The process of improving the processes within the laboratory creates a need to review the appropriateness of the organizational structure in a way that encourages people to continue their effort towards improvement.

Implementing TQM
Organizational Structure
- Incremental improvement of processes
- Review of the appropriateness of the organizational structure:
  - Laboratory management processes
  - Methods of resource allocation
  - Administrative support processes
  - Human environment
  - Training for all laboratory members
  - Laboratory processes and procedures

Such changes to management structure include
- changes to the management processes of the laboratory such as reward or payment, recognition and strategic planning,
- changes to the methods of resource allocation,
- changes to the administrative support processes such as secretarial, clerical and purchasing
- building and maintaining a human environment that allows the laboratory personnel to improve quality continually, based on mutual trust and collaboration,
- enhancing training for all laboratory members and
- improving laboratory processes and procedures.

Slide 17

The organization must consider all management and operational activities as processes and measure their efficiency and effectiveness. The process owner and process customer of each process as well as management responsible for the processes must be clearly defined. Within each process the responsibilities of management and process owner must be conveyed to the personnel that perform such processes.

Implementing TQM
Process Management Concept
- Process owner and process customer
- Responsibilities of management and process owners:
  - Purpose of each process
  - Customers of each process
  - Needs and expectations of customers
  - Needs and expectations of the process owners
  - Performance standards of processes
  - Measurement of process performance
  - Improvement opportunities

The laboratory management is responsible for defining and agreeing the purpose of each one of the laboratory activities and must explain to individuals the relationship of each process to the overall objectives that are rarely known to the technical people.

The purpose of each process and the performance standards for the key activities of the process must be defined and agreed with those that are interested in the results of each such process.

The needs and expectations of customers and process owners must be identified. The performance of each process is measured and monitored and opportunities for improvement are sought, the aim being the continual improvement.

Slide 18

The performance of all key laboratory functions and processes must be assessed based on appropriate perform-ance standards.

The performance of functions and processes should be based on key attributes and the process efficiency should be based on well defined indicators.

Measurement of customer satisfaction (internal and external customers) as

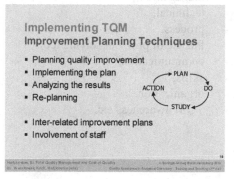

well as financial factors should also be considered. This may be based on customer surveys. Such surveys should include, and not limited to laboratory services, customer complaints, reject and reanalysis costs, measures of the time to complete an analysis or delay time to release a batch etc.

Slide 19

Planning quality improvement, implementing the plan, analysing the results and re-planning is a continuous cycle (Plan-Do-Check-Act). Inter-related improvement plans are deployed at all levels of the laboratory (corporate, department, process, indi-vidual). It is essential to note that the commitment of laboratory personnel is maximized when such personnel is actively involved in the generation and deployment of the improvement plans.

Slide 20

All laboratory personnel, including the highest levels of management, should receive training and education. Such education will enable them to carry out more efficiently and more effectively their individual processes and functions, to be aware of the relationship of the various laboratory processes, to understand the importance of customer satisfaction and the corporate laboratory objectives and to be able to contribute effectively to the continuous improvement programme.

Implementing TQM
Training

- Why training
  - To perform individual process
  - To be aware of the relationship with other processes
  - To understand the significance of our role and the part we play in customer satisfaction and business objectives
  - To contribute to the continual improvement programs

Slide 21

Training programs are important in constructing and maintaining a quality improvement culture in the laboratory. Such programs should be extended to all members of the laboratory so that they are able to understand the quality principles and practices and the appropriate methods and techniques for quality improvement. Subjects for training can include
- management,
- technical,
- process,
- the use of problem solving tools
- communication,
- skills,
- organization,
- TQM awareness etc.

Implementing TQM
Training

- Environment for quality improvement
- Training to all laboratory personnel
  - Quality principles and practices
  - Quality improvement methods
- Subjects for training
  - Management, technical, process, problem solving tools, communication, skills, organization, awareness etc

Slide 22

Quality improvement is a continuous activity, aiming to ever higher levels of process effectiveness and efficiency. Situations requiring improvements may arise from different studies and surveys, from systematic monitoring of the laboratory activity, from analyses of various measurement results and statistics etc.

A quality driven laboratory is constantly seeking the opportunities for improvement arising when problems are encountered and during problem solving processes. Therefore, problem solving techniques should be applied to all areas of the business and all individuals and groups should be encouraged to use them. For the best use of available resources priorities for improvement should be reviewed before action is taken.

Quality Improvement

- Situations requiring improvement
  - High quality costs
  - Customer complaints
  - Health and safety considerations
- Problem solving techniques
  - Identify opportunities for improvement
  - Apply to all areas of the business
- Review priorities of improvement before action

Slide 23

The management shall motivate its members to undertake quality improvement projects or activities in a consistent and disciplined series of steps based on data collection and analysis.

The quality improvement process is initiated and its need, scope and importance are clearly defined.

Investigation of possible areas for improvement increases the understanding of the nature of the process to be improved.

The data of the process to be improved are analyzed and a cause and effect relationship is formulated.

Alternative proposals for improvement actions are developed and evaluated by the members of the organization involved in such processes.

The improvement is confirmed by analyzing more data and the process is introduced into the system.

Quality Improvement
A Methodology for Quality Improvement

- Involve the whole organization
- Initiate quality improvement projects or activities
- Investigate possible areas for improvement
- Establish cause and effect relationship
- Take improvement action
- Confirm the improvement
- Sustain the gains

Slide 24

- The quality improvement project or action starts with the identification of an improvement opportunity which is based on measures of quality losses and/or on comparisons against organizations recognized as leaders in a particular field. The opportunity for improvement is defined and the process involved is evaluated by analyzing data and facts. The objectives for improvement are introduced.

> **Quality Improvement**
> **Improvement Process**
>
> - Identify
> - Evaluate
> - Plan
> - Execute
> - Check
> - Amend

- A set of processes is introduced to achieve the improvement in question. Data from such processes are analyzed and evaluated and the causes requiring improvement are identified. The objectives for improvement are also identified.
- Possible solutions are devised and the preferred one is selected. The implementation of the solution is then planned.
- The plan is implemented.
- The results are monitored, reviewed and evaluated.
- The process is repeated if solution is not achieving its objectives.

Slide 25

The problems encountered in the laboratory operations are considered as opportunities for improvement. The problem solving process is an excellent tool for people to understand how a process performs and how the laboratory operates.

> **Quality Improvement**
> **Problem Solving Process**
>
> - Identify subjects for improvement
> - Prioritize
> - Analyze causes of problem
> - Collect data for analysis
> - Assess alternative solutions for actions
> - Select the optimum solution for action

The problem solving process starts with the identification of subjects for improvement. Such subjects are then placed in order of priority. The causes of the problem are identified and analyzed and then, data are collected for analysis and evaluation. Various solutions are assessed and the optimal one is selected.

Slide 26

Another, similar approach to the problem solving process, is the one shown in this slide. Here, the following questions have to be answered:

- Which problem should I address? If there are several, how do I choose the most important one?
- How do I accurately and completely describe the problem?
- What are the different causes of the problem, and which causes are the important ones to solve right away?
- What are the possible solutions for solving the problem?
- How do I make sure the solutions are implemented correctly and effectively?
- How did the solutions work? What needs to be changed?

**Quality Improvement
Problem Solving Process**

- Identifying the problem
- Describing the problem
- Analyzing the problem
- Planning the solutions
- Implementing the solutions
- Monitoring/evaluating the solutions

Slide 27

Quality improvement needs first a well structured organization, both horizontally and vertically.
Within the organizational hierarchy, responsibilities for quality improvement include

- management processes (i.e. provision for the policy, the strategy and the major quality improvement goals, provision of resources, education and training etc.)
- work processes (i.e. how the various operations are performed),
- how measurements are done and how quality losses are calculated,
- how secretarial, budgeting and purchasing processes are performed and
- how an environment that empowers, enables, and charges all members of the organization to continuously improve quality is built and maintained.

**Managing for Quality Improvement
Organizing for Quality Improvement**

- Responsibilities for quality improvement
  - Within the organizational hierarchy
    - Management processes
    - Work processes
    - Measurement of the reduction of quality losses
    - Administrative support processes
    - Building of an environment for quality improvement
  - Within the processes that flow across organizational boundaries

Slide 28

Within the processes that flow across organizational boundaries the responsibilities for quality improvement are shown in this slide.

Managing for Quality Improvement
Organizing for Quality Improvement

- Responsibilities for quality improvement
  - Within the organizational hierarchy
  - Within the processes that flow across organizational boundaries
    - Definition of the purpose of each process
    - Communication among departments
    - Identification of internal and external customers
    - Determination of their needs and expectations
    - Searching process improvement opportunities

Slide 29

The laboratory needs to plan for quality improvements.
The planning process should provide for strategic guidance and directions for meeting quality improvement goals and implementing quality policy. Quality improvement activities that are employed to achieve the quality improvement plans should fit in the overall goals and business objectives. The planning process must have inputs from all levels of the laboratory, from reviews of achieved results, and from suppliers and customers.

Managing for Quality Improvement
Planning for Quality Improvement

- Set quality improvement goals
- Address the most important quality losses
- Involvement of everyone
- Inputs from all, from reviews, from suppliers from customers
- Focus on newly identified opportunities and where there is insufficient progress
- Implement quality improvement plans

Slide 30

Quality improvement as a term denotes transition from a stage of inferior quality to a stage of superior quality. This change should always be measured in order to prove that the transition to the better stage is successful. The measurements should relate to quality losses associated with
-    customer satisfaction,
-    process efficiencies, and
-    losses sustained by society.
This slide shows the basis for measurements associated with customer satisfaction.

Managing for Quality Improvement
Measuring Quality Improvement

- Measure of quality losses
  - Associated with customer satisfaction
    - Surveys of current and potential customers
    - Surveys of competing products and services
    - Changes in revenues
    - Inspections
    - Customer complaints
  - Associated with process efficiency
  - Sustained by society

Slide 31

The basis of measurements of quality losses associated with process efficiency.

Managing for Quality Improvement
Measuring Quality Improvement

- Measure of quality losses
  - Associated with customer satisfaction
  - Associated with process efficiency
    - Labour, capital and material utilization
    - Unsatisfactory process output
    - Waiting times
    - Delivery performance
    - Size of inventories
  - Sustained by society

Slide 32

The basis of measurements of quality losses sustained by society.

Managing for Quality Improvement
Measuring Quality Improvement

- Measure of quality losses
  - Associated with customer satisfaction
  - Associated with process efficiency
  - Sustained by society
    - Failure to realize human potential (surveys of employee satisfaction)
    - Damage caused by pollution and disposal of waste and depletion of scarce resources

Slide 33

Trends displayed by measurements should be interpreted statistically and numerical targets should be established. The measurements are reported and reviewed and trends are evaluated. The cost of measurement is significant and should always be considered.

Managing for Quality Improvement
Measuring Quality Improvement

- Statistical interpretation of trends
- Establish and meet numerical targets
- Measure and track trends
- Report and review measures
- Measure the cost of measurement

Slide 34

A variety of data, either numerical or non-numerical can be gathered in a systematic manner for a clear and objective picture of the facts. This is the *data collection form*. Such forms are very common in the analytical laboratories and are used to collect and record data and facilitate comparisons.

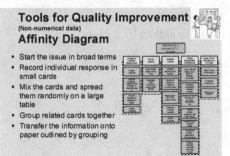

Slide 35

This organizes a large number of opinions, ideas or concerns into group-ings based on the natural relationships that exist among them. It is often used to organize ideas generated by brain-storming. The procedure of applying an affinity diagram consists of recording the individual ideas in small cards and then grouping related cards together. The information from the cards is then transferred onto paper outlined by groupings.

The affinity diagram was developed for discovering meaningful groups of ideas within a raw list. In doing so, it is important to let the groupings emerge naturally, using the right side of the brain, rather than according to preordained categories.

Slide 36

Benchmarking is used to compare an organization's activity against that of a recognized leader in the market. This will identify opportunities for quality improvement and will lead to competi-tive advantage in the market place.

Slide 37

Brainstorming is an excellent technique for tapping the creative thinking of a team quickly to generate, clarify, and evaluate a sizable list of ideas, problems, issues etc. The team leader stimulates the team to create numerous ideas irrespective of their quality, which are then clarified and evaluated. During the generation phase the team leader reviews the rules for brainstorming and the team members express their ideas which are then clarified so that everyone understands them. The ideas are evaluated to extract useful information and make appropriate decisions.

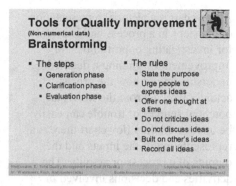

Slide 38

A cause and effect diagram (sometimes known as the *"Ishikawa"* or the *"fishbone diagram"*) represents the relationships between a given effect and its potential causes. The cause and effect analysis relates the interactions among the factors affecting a process. The cause and effect diagram is widely used when identifying the effects on a result, including a chemical analysis result. It is used for example in measurement uncertainty to analyse the uncertainty sources. A cause and effect diagram describes a relationship between variables. The undesirable outcome is shown as an effect, and related causes are shown as leading to, or potentially leading to, this effect.

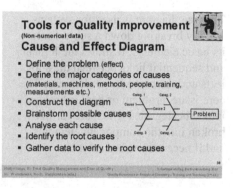

Slide 39

A flowchart is a pictorial representation of the steps in a process and is useful for investigating opportunities for improvement by gaining a detailed understanding of how the process actually works. From the flowchart, potential sources of trouble can easily be uncovered. In a flowchart there is a representation of the inputs and the outputs of a process or activity. The activities and decisions involved in converting an input to an output are shown within the appropriate flowchart symbols.

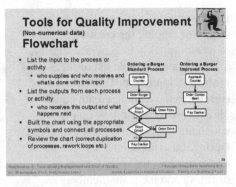

Slide 40

The tree diagram is used for systematically breaking down a subject into its basic elements. It shows the logical and sequential links between the subjects and the component elements. The core issue, problem or goal is broken into subcategories which are further broken into the component elements and if necessary sub-elements.

Slide 41

A control chart displays statistically determined upper and lower limits drawn on either side of a process average. This chart shows if the collected data are within upper and lower limits previously determined through statistical calculations of raw data from earlier trials.
The construction of a control chart is based on statistical principles and statistical distributions, particularly the normal distribution. When used in conjunction with a manufacturing process, such

charts can indicate trends and signal when a process is out of control. The centre line of a control chart represents an estimate of the process mean; the upper and lower critical limits are also indicated. The process results are monitored over time and should remain within the control limits; if they do not, an investigation is conducted for the causes and corrective action taken. A control chart helps determine variability so it can be reduced as much as is economically justifiable.

In preparing a control chart, the mean upper control limit (UCL) and lower control limit (LCL) of an approved process and its data are calculated. A control chart with mean UCL and LCL with no data points is created; data points are added as they are statistically calculated from the raw data. (See also the chapter on control charts)

Slide 42

The histogram is a visual representation of the distribution of variable data. The information is represented by a series of equal-width columns of varying heights. Column heights represent the number of observations. The natural tendency of data is to fall towards the centre of the distribution, with progressively fewer towards the extremes.

Tools for Quality Improvement
(Numerical data)
Histogram

- Collect data
- Arrange in ascending order
- Determine the range of the data
- Determine the width of each class interval (column)
- Put class interval in the X-axis
- Put frequency scale on the Y-axis
- Draw the height of each column

After the raw data are collected, they are grouped in value and frequency and plotted in a graphical form. A histogram's shape shows the nature of the distribution of the data, as well as central tendency (average) and variability. Specification limits can be used to display the capability of the process.

Slide 43

Causes can be ranked from most to least significant. The *Pareto diagram* is based on the *Pareto principle*, which states that just a few of the causes account for most of the effect. Quality experts often refer to the principle as the 80-20 rule; that is, 80% of the problems are caused by 20% of the potential sources. By distinguishing

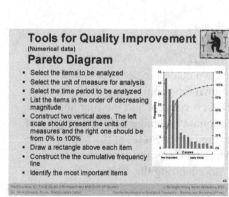

Tools for Quality Improvement
(Numerical data)
Pareto Diagram

- Select the items to be analyzed
- Select the unit of measure for analysis
- Select the time period to be analyzed
- List the items in the order of decreasing magnitude
- Construct two vertical axes. The left scale should present the units of measures and the right one should be from 0% to 100%
- Draw a rectangle above each item
- Construct the the cumulative frequency line
- Identify the most important items

the more important causes from the less significant ones the greatest improvement will be obtained with the least effort. The Pareto diagram displays, in decreasing order, the relative contribution of each cause (based for example on the number of occurrences, the cost associated with each cause etc.) to the total problem.

Slide 44

A scatter diagram shows how two variables are related and is thus used to test for cause and effect relationships. It cannot prove that one variable causes the change in the other, only that a relationship exists and how strong it is. In a scatter diagram, the horizontal (x) axis represents the measurement values of one variable, and the vertical (y) axis represents the measurements of the second variable.

Tools for Quality Improvement
(Numerical data)
Scatter Diagram

- Collect paired data
- Graduate x and y axes
- Plot the data
- Label the axes
- Examine the pattern

Slide 45

W.A. Shewhart, F.W. Taylor, W.E. Deming, J.M. Juran, K. Ishikawa, P.B. Crosby, A.V. Feigenbaum, T.J. Peters, G. Taguchi, T. Ohno, S. Shingo are considered to be the most important persons related to quality in the last 100 years.

Quality Gurus

- Philip B. Crosby
- William E. Deming
- Joseph M. Juran
- Armand V. Feigenbaum
- Kauro Ishikawa
- Tom J. Peters

Slide 46

Philip B. Crosby was an internationally known quality expert. He is best known for popularizing the *Zero Defects* concept and the four absolutes for quality management that originated in the United States at the Martin Marietta Corporation where *Crosby* worked during the 1960s.
Philip Crosby used a disciplined and highly structured approach, which is

Quality Gurus
Philip B. Crosby

The four absolutes of Quality Management

1. Quality equals conformance to requirement
2. Prevention causes quality
3. Zero defects
4. The measurement of quality is the price of non-conformance

not solely product-oriented, but is based entirely on "*prevention*" and is readily applicable to any enterprise. A new management quality commitment and culture program is achieved through "*the four absolutes*" of *Crosby*.

Slide 47

A well structured approach, which demonstrates how the culture can be changed and a process improved, is provided through the *Crosby's* 14-step quality improvement process.

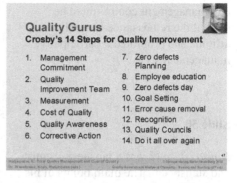

Slide 48

William E. Deming was an American statistician who was distinguished in Japan and only after 30 years did America and the western countries realize his prophetic philosophy. Through his philosophy, encoded into his 14 points of management obliga-tions and management commitment, he removed the major roadblocks to quality improvement and he started the renaissance in quality attitude and promoted a participative management style. *Deming* is famous for his quality circle which prescribes good management as a cyclical process which follows the order of Plan-Do-Check-Act.

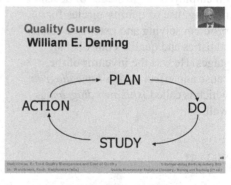

Slide 49

Deming's philosophy is as much about management style and leadership as the practice of quality itself. Through his 14 point of management obligations and management commitment he started the renaissance in quality attitude and promoted a participative management style.

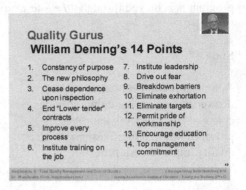

Slide 50

Kauro Ishikawa was a Japanese chemist and statistician. The main points of his methodology include involvement of all employees in all stages of decision making, use of quality circles for problem solving and extensive use of statistics and quality control at all stages. He was the inventor of the cause and effect or *fish-bone diagram*, which is called *Ishikawa diagram* as well.

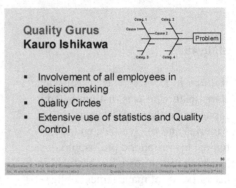

Slide 51

Joseph M. Juran was one of the most well known gurus in the quality scene, worldwide. He first visited Japan in 1954 and started conducting seminars explaining to top managers the role they had to play in promoting QC activities.

Juran developed the idea of the *Quality Trilogy:* Quality Planning, Quality Improvement and Quality Control. These three aspects of company-wide

strategic quality planning are further broken down in *Juran's "Quality Planning Road Map"*, into the following key elements.

Quality Planning
- Identify who are the customers.
- Determine the needs of those customers.
- Translate those needs into our language.
- Develop a product that can respond to those needs.
- Optimize the product features so as to meet our needs and customer needs.

Quality Improvement
- Develop a process which is able to produce the product.
- Optimize the process.

Quality Control
- Prove that the process can produce the product under operating conditions.
- Transfer the process to operations.

Slide 52

Juran stressed that any organization produces and distributes its products through a series of specialized activities carried out by specialized departments. These activities (actions) are depicted by the spiral of progress in quality. The spiral shows actions necessary before a product or service can be introduced to the market. Each specialized department in the spiral (e.g., customer service, marketing, purchasing) is given the responsibility to carry out its assigned special function. In addition, each specialized department is also assigned a share of the responsibility of carrying out certain company-wide functions such as human relations, finance, and quality. Quality results from the interrelationship of all departments within the spiral. *Juran* talked about the quality function to describe activities through which the departments around the spiral can attain quality.

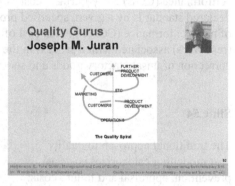

Slide 53

Cost of quality is the total cost of ensuring quality in products and services (including analytical activities). This cost enables the organization to measure the effect of quality on current practices. Total quality management requires the management of processes, not just the outputs. Every person within the organization contributes to and operates within a process, and every process should have an identified process owner who is responsible for the effectiveness of the process. Process cost is the total cost of conformance and cost of non-conformance. Cost of conformance (COC) is the intrinsic cost of providing products or services to declared standards by a given, specified process in a fully effective manner. Cost of non-conformance (CONC) is the cost of wasted time, materials and capacity (resources) associated with a process in the production, dispatch, receipt, and correction of unsatisfactory goods and services.

Cost of Quality

- Measurement of the effect of quality
- Management of processes
- Involvement of all individuals to the processes
- Process owner
- Cost of conformance
- Cost of non conformance

Slide 54

The traditional approach to quality cost modelling categorizes costs as prevention, appraisal and failure costs. Prevention and appraisal costs are the costs of assuring quality, and internal and external costs are the failure and scrap costs. Added together, these produce the total costs.

The cost of prevention is that of designing a process, e.g. an analytical method, right in the first place, training laboratory, organizing the laboratory so that work is done efficiently and effectively and improvements can be made all the time.

The cost of appraisal in a laboratory is about measuring the laboratory performance, monitoring of the laboratory processes, doing internal quality control, internal and external laboratory audits and so on.

Cost of Quality

- Costs of control
  - prevention costs
    - The cost of training people, organizing business etc.
  - appraisal costs
    - Quality control (analyzing, testing, measuring etc.)

Slide 55

Internal failure cost is the cost when you discover, before release a laboratory report or an analysis certificate that something has gone wrong. External failure cost is where an analysis result or certificate reach the customer before you discover that something has gone wrong.
The lost opportunity cost is the cost when we do not proceed to the routine laboratory activities as a result of

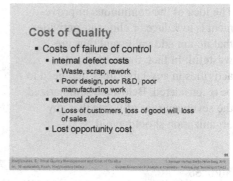

other, non productive activities that must take place in the laboratory. In other words, this cost arises when we are dealing with trivial activities that prevent us from doing more important activities. The cost of this lost opportunity is probably the highest of the costs of quality.

Slide 56

The cost for appraisal and prevention per unit of a good quality product gets higher and higher when the desired defects are minimal, i.e. when we tend to a 100% good quality product. On the other hand the costs of internal and external quality becomes lower and lower when we tend to the production of 100% good product. The minimum cost of quality is found at the lower point of the sum of the two graphs of

the cost of appraisal and prevention and that of the internal and external failure costs. A "*quality product*" in a laboratory could be for example an analysis performed on time with optimal quantities of reagents.

Slide 57

The idea of the continuous improvement is to reduce or eliminate activities that do not add value and thus are wasteful. In fact, there are many activities in any laboratory that need to be reconsidered. Below are summarized the seven sources of waste that any organization should avoid.

Cost of Quality
The Seven Sources of Waste

- Overproduction
- Defective products
- Waiting lines and delays
- Stocks of intermediaries/semi-finished products
- Transportation
- Ineffective procedures
- Ineffective movements or actions

Slide 58

This is the waste coming from the production of more units than demanded. The optimum number of products to be produced must balance the demand, including that in high season periods, the cost of holding the stocks in the warehouse and the cost of setting up the production to produce one lot of the product. In terms of an analytical laboratory, overproduction could be interpreted as any activity

Cost of Quality
The Seven Sources of Wastes

- Overproduction

that is not necessary for customer service or for adding value to the experience and knowledge of the laboratory. There is no need to analyse too many samples unless we have reasons to do so. Furthermore, there is no need to perform more analyses than necessary. Too many analytical results are a waste. For example, the optimum number of replicates must balance the need for statistical evaluation.

Slide 59

This is the waste coming from the production of defective products. The manufacturing unit must set up a system to produce the right products first time in order to avoid the production of defective products, which are either discarded or reprocessed, incurring a significant cost to the organization. A "*defective product*" for an analytical laboratory is considered any activity or analytical results without any practical use to the laboratory.

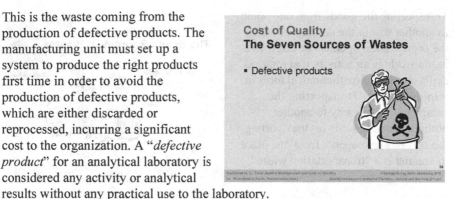

Slide 60

It is a waste of time and money when projects remain too long in waiting lines or there are delays in executing specific tasks. It is the responsibility of the management to design a proper plan in order to distribute the planned actions evenly. Imagine a series of vials in the tray waiting to be processed! Is it necessary for all such samples to be processed? Are they in the correct order? If not this is a waste of time or reagents.

Slide 61

Keeping stocks (solvents, materials etc.) enable the laboratory to be responsive to the needs of the customer and in the execution of the analytical work at the right time. However keeping stocks in quantities beyond the minimum can be very costly.

Slide 62

Transporting the goods from one point to another within the country in which the organization operates or elsewhere in the world is an activity that requires skills and use of mathematical tools to minimise costs. Transporting the samples from one city to another without sound reasons or transporting the solvents or reagents from one place to another is a "transportation waste".

Slide 63

The procedures that are followed for the optimum performance of a process or action are very important for more effective and efficient operations. Therefore, keep your processes and procedures as unambiguous as you can, but make them as simple as necessary.

Slide 64

All activities within the laboratory must be well structured and organized to avoid ineffective movements or actions. This ineffectiveness can be corrected with the proper layout of instruments and work areas in the laboratory and a proper laboratory activity plan.

Slide 65

Total Quality Management is the applied philosophy that makes businesses established as leaders in the market. The management committed to the spirit of TQM ensures that the organization does not just survive but achieves excellence. The modern business environment is subject to rapid changes that take place world-wide and organizations seeking excellence should manage the change effectively and succeed by incremental and continual improvement. Problems that arise in the business activity should be seen as opportunities for improvement and the problem solving process as the route to Ithaca i.e. the route to excellence.

Summary

- Applied philosophy
- Management commitment
- Achievement of excellence
- Management of change
- Continual improvement
- Opportunities for improvement
- Tools for improvement

Slide 66

Where to Get More Information

- http://www.philipcrosby.com
- http://www.isixsigma.com
- http://www.deming.org/
- http://www.juran.com/
- BS 7850-1:1992 Total Quality Management, Part 1: Guide to management principles
- BS 7850-2:1994 Total Quality Management, Part 2: Guide to quality improvement methods
- BS 6143-1:1992 Guide to the economics of Quality. Part 1: Process cost model
- BS 6143-2:1990 Guide to the economics of Quality. Part 1: Prevention, appraisal and failure model

Bibliography:

Arvanitoyiannis I, Hadjicostas E (2001) Quality Assurance and Safety Guide for the Food and Drink Industry. CIHEAM/Mediterranean Agronomic Institute of Chania / European Commission MEDA

BS 6143-1:1992 Guide to the economics of Quality. Part 1: Process cost model

BS 6143-2:1990 Guide to the economics of Quality. Part 1: Prevention, appraisal and failure model

BS 7850-1:1992 Total Quality Management, Part 1: Guide to management principles

BS 7850-2:1994 Total Quality Management, Part 2: Guide to quality improvement methods

Friedman FB (1995) Practice Guide to Environmental Management. 6th edition; Environmental Law Institute, Washington

Hradesky JL (1987) Productivity and Quality Improvement, A Practice Guide of Implementing Statistical Process Control; McGraw Hill

Krajewski JL, Ritzman PL (1999) Operations Management, Strategy and Analysis. 5th edition; Addison-Wesley

Mullins E (1994) Introduction to control charts in the analytical laboratory, Analyst 119, 369-375

Pierce FD (1995) Total Quality for safety and health professionals. Government Institutes USA

Pyzdek T (1990) Guide to SPC. ASQC Quality Press, Quality publishing

7 Quality Manual

Michael Koch

The quality manual is the "heart" of every management system related to quality. Quality assurance in analytical laboratories is most frequently linked with ISO/IEC 17025, which lists the standard requirements for a quality manual. In this chapter examples are used to demonstrate, how these requirements can be met. But, certainly, there are many other ways to do this.

Because of the importance of ISO/IEC 17025 it is not possible to deal with this subject without referring to this standard. In the slides all the citations from ISO/IEC 17025 are written in italics.

Slide 1

ISO/IEC 17025 is the most important standard for quality assurance in analytical laboratories. All the quality assurance measures must be described in a quality manual.

Quality Manual – Why?

- ISO/IEC 17025 states the necessity for an accredited laboratory to have a management system related to quality with certain requirements as well as procedures to keep documents under control

- The quality manual is the keystone of the documentation of a quality management system.

Slide 2

This presentation is structured in two parts. One deals with the contents of a quality manual and the other with the necessary measures for document control.

Quality Manual

- Contents

- Document Control

B.W. Wenclawiak et al. (eds.), *Quality Assurance in Analytical Chemistry: Training and Teaching*, DOI 10.1007/978-3-642-13609-2_7, © Springer-Verlag Berlin Heidelberg 2010

Slide 3

In chapter 4.2.1 of ISO/IEC 17025 the necessity for a management system and for documentation of policies, systems, procedures etc. is stated. It is important that the system is appropriate for the laboratories activities (fit for the purpose!). The system and documentation only make sense, if they provide feedback to the personnel. Therefore the documentation has to be communicated to, understood by, available to and implemented by the appropriate personnel.

> **Quality Manual – Contents**
> **ISO/IEC 17025 – 4.2.1**
>
> • The laboratory shall establish, implement and maintain a management system appropriate to the scope of its activities.
> • The laboratory shall document its policies, systems, programmes, procedures and instructions to the extent necessary to assure the quality of the test and/or calibration results.
> • The system's documentation shall be communicated to, understood by, available to, and implemented by the appropriate personnel.

Slide 4

Clause 4.2.2 of ISO/IEC 17025 requires that a quality policy statement be issued by the top management and that the quality system policy and objectives are defined and documented in the quality manual.

> **Quality Manual – Contents**
> **ISO/IEC 17025 – 4.2.2**
>
> • The laboratory's management system policies related to quality, including a quality policy statement, shall be defined in a quality manual (however named).
> • The overall objectives shall be established, and shall be reviewed during management review.
> • The quality policy statement shall be issued under the authority of top management.

Slide 5

This chapter also lists the necessary contents of a quality manual:
- the commitment to good professional practice
- a statement of the standard of service
- the purpose of the system to ensure quality

> **Quality Manual – Contents**
> **ISO/IEC 17025 – 4.2.2 (followed)**
>
> • It shall include at least the following:
>   • the laboratory management's commitment to good professional practice and to the quality of its testing and calibration in servicing its customers;
>   • the management's statement of the laboratory's standard of service;
>   • the purpose of the management system related to quality;

Slide 6

(Continued content):
-   commitment of personnel to implement of the policies and procedures and to act according to the requirements of the system.
-   the management's commitment to compliance with ISO/IEC 17025 and to continual improvement

Quality Manual – Contents
ISO/IEC 17025 – 4.2.2 (followed)

- a requirement that all personnel concerned with testing and calibration activities within the laboratory familiarize themselves with the quality documentation and implement the policies and procedures in their work; and
- the laboratory management's commitment to comply with this International Standard and to continually improve the effectiveness of the management system

Slide 7

The quality manual also has to include or has to make reference to all supporting and technical procedures. The quality manual therefore is the central document of the quality management system.
The structure of the documentation also has to be outlined in the quality manual.

Quality Manual – Contents
ISO/IEC 17025 – 4.2.5

- The quality manual shall include or make reference to the supporting procedures including technical procedures.
- It shall outline the structure of the documentation used in the management system.

Slide 8

The roles and responsibilities of the technical and quality management have to be described separately in the quality manual.

Quality Manual – Contents
ISO/IEC 17025 – 4.2.6

- The roles and responsibilities of technical management and the quality manager, including their responsibility for ensuring compliance with this International Standard, shall be defined in the quality manual.

Slide 9

As mentioned above the quality policy is one of the most important parts of the quality manual. The whole system should only reflect and realize the quality statements and requirements in the daily work. The policy shall be authorized by the top management, because this statement has to reflect clearly, how crucial quality aspects are in the work of the laboratory.

Quality Policy

- The quality policy is one of the essential things in the management system.
  - the whole system should only reflect the quality policy in the daily work and put the policy into concrete terms
- the policy shall be issued under the authority of the top management
  - It is very important to have a clear statement, how crucial quality is in the work of the laboratory

Slide 10

The quality policy has to be very specific for each laboratory, because it depends on the tasks and the clients of the laboratory and eventually on its role in a larger organization.
Therefore no recipe can be given for the formulation.

Quality Policy

- Quality policy is very specific for each laboratory in the framework of its tasks, its role in a larger organization and its relationship to its customers.
- There is no recipe for the formulation

Slide 11

The quality manual should cover all the aspects given in this slide.
The statement of independence from influences that may adversely affect the quality of its work is very important and special care has to be taken to identify possible sources of these influences.

Presentation of the Laboratory
- This chapter should contain:
  - Address
  - Phone, Fax, e-mail
  - Internet-Website
  - Ownership
  - Year of foundation
  - History
  - Bank account
  - Memberships in associations and organisations
  - Statement of independence from influences that may adversely affect the quality of the work
  - Cooperation with other laboratories and organizations
  - ...

Slide 12

The organization and laboratory management have to be described in another chapter of the quality manual. This must include the names of persons responsible for the commercial and for the technical management and their deputies. The qualification of the technical manager and his/her deputy must be described in order to show that he/she is able to fulfil the necessary requirements. The whole organisation should be described in an organization chart.

Organization and Management

- Names of persons responsible for commercial management
- Names of persons responsible for technical management
  - including a short description of the qualification of the technical manager and his deputy
- Organization chart

Slide 13

The organization chart has to reflect the structure of the laboratory. It has to show: *who is responsible to whom?* and *who is who's boss?*

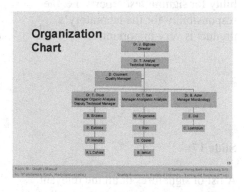

Organization Chart

Slide 14

The laboratory must have a quality manager. He/she has to be named in the quality manual together with his/her responsibilities. The quality manager must have direct access to the highest level of management responsible for the laboratory in order to have the possibility to indicate on problems that could affect the quality of the work.

Quality Manager

- The laboratory must have a quality manager (and if possible a deputy)
- The name and responsibilities/duties have to be stated here
- The quality manager must have direct access to the highest level of management at which decisions are made on laboratory policy or resources

Slide 15

The chapter "*staff*" of the quality manual should state:
- the number of employees
- the persons responsible for sub-divisions
- and make reference
- where information for all technical staff for the items listed is documented

Staff
- Number of employees
- Names of persons responsible for subdivisions
- Reference, where information for all technical staff is documented concerning
  - Relevant authorizations
  - Competence
  - Educational and professional qualifications
  - Training (in the past and plans for the future)
  - Skills
  - Experience

Slide 16

The responsibilities listed here have to be allocated. Especially the responsibility for signing test reports, i.e. the responsibility for the laboratory's product is very important.

Allocation of Responsibilities
- For signing contracts
- For signing test reports
- For acquisition
- For procurement

Slide 17

A list of signatures of all the relevant staff has to be included in the quality manual in order to identify them in the documentation of the quality system.

List of Signatures
- For all relevant staff

Slide 18

A quality manual does not make any sense, if the staff do not have access to it. However, it is imperative that such access has to be controlled. It has to be ensured that all copies in use are current authorized versions. Every authorized copy must therefore be marked as such and there must be a list with a clear identification of every authorized copy and where it is located. If the quality manual is changed, it has to be ensured that all authorized copies are changed as well. The quality manual has to be reviewed regularly to see if it is still up to date. The management of access to and review of the quality manual are one of the duties of the quality manager.

Administration, Access and Review of the Quality Manual

- All staff must have access to the quality manual
- This access must be managed
- All copies must be current versions
- The quality manual must be regularly reviewed
- All these things are the duties of the quality manager

Slide 19

All operations that are done regularly in the laboratory, especially the analyses, have to be described in standard operation procedures (SOP). It is useful to describe the standard operations in separate documents and to include only a list of all SOP's in the quality manual.

Standard Operation Procedures

- There have to be standard operation procedures for all relevant procedures
- It is useful to have the SOPs in separate documents
- The quality manual should contain a list of available SOPs

Slide 20

The next two slides describe the typical content of a SOP for a test method:
-   a general description to give an overview
-   the identification of any underlying written standard (like ISO, ASTM, BSI, DIN etc.)
-   special requirements for sampling and conservation (or a link to a separate SOP for that)

Standard Operation Procedure – Content - I

- General description of the method
- Underlying standard
- Sampling and conservation
- Range of application
- Interferences
- Necessary equipment
- Chemicals (purity, where to buy)
- Measurement

- the range of application for this analysis (e.g. drinking water, waste, sediments, contaminated soils)
- interferences due to other ingredients
- necessary equipment and chemicals for the analysis
- detailed description of the measurement process

Slide 21

SOP content (continued):
- requirements for calibration: e.g. calibration standards, external/ internal/standard addition, calibration intervals...
- evaluation: e.g. use of an internal standard, ...
- use of control charts (see chapter 13)
- use of reference materials (see chapter 14)
- presentation of results: how many significant digits? which unit?
- limit of detection, limit of qunatification, measurement uncertainty (see chapter 9 and 12)
- responsibilities: who is responsible? Who should be called in case of problems?

Standard Operation Procedure – Content - II

- Calibration
- Evaluation
- Control charts
- Other quality assurance measures
- Use of reference materials
- Presentation of results
- Limit of detection, limit of quantification, measurement uncertainty
- Responsibilities

Slide 22

Many of the items mentioned above are described in detail in written standards, but not all of them. The details in this slide are laboratory specific and must therefore be described in a SOP.

The extent of a SOP depends also on the qualification of the staff. If this is high, the SOP can be much shorter compared to an SOP for low qualified staff.

Standard Operation Procedure – Standardised Methods

- A standard method cannot cover the laboratory specific details
  - Equipment
  - Supplier
  - Trained staff
  - Responsibilities
  - QA measures
  - Limit of detection, ...
- These details have to be documented
- The extent of the SOP depends on the qualification of the staff

Slide 23

Laboratory internal guidelines for general operations like the calculation of detection limits or the measurement uncertainty should be included in the quality manual or reference made to further documentation.

Other Guidelines

- If the laboratory has other guidelines, they should be included or referenced in the quality manual
  - For calibration
  - For calculation of detection limits
  - For calculation and construction of control charts
  - For ...

Slide 24

Reference materials have to be used, whenever possible. The laboratory has to keep a list of the reference materials in use. This list has to be included or referenced in the quality manual.

Reference Materials

- The laboratory should keep a list of reference materials used
- This list has to be in the quality manual
- Or a note where the RM's are listed

Slide 25

The accommodation and the environmental conditions in the laboratory must fulfil certain requirements. Such requirements have to be described in the quality manual. Usually the rooms are listed and a floor plan is added.

Accommodation and Environmental Conditions

- ISO/IEC 17025 contains requirements for accommodation and environmental conditions
- Their fulfilment must be described in the quality manual
- Description of the rooms
- Floor plan

Slide 26

Each piece of equipment, which is relevant to the tests performed, must have a log book. All changes, calibrations, maintenance etc. must be recorded there. Usually these records are not kept in the quality manual. But a list of all existing log books has to be included. It might be useful to have a guideline on what has to be included in these records.

Equipment

- There must be records on each item of equipment and its software significant to tests and/or calibration performed
- Usually on separate documents
- List of all records
- (Guideline, what has to be recorded)

Slide 27

No equipment lasts forever. Therefore, if it is defective or its operation is out of the specified limits, it has to be marked as "out of service". There must be regulations in the quality manual on how this should be done, what must be arranged, who has to be informed and who is responsible for looking for a replacement.

Defect and Incorrect Working Test Equipment

- Equipment that has been shown to be defective or outside specified limits must be taken out of service
- There must be regulations in the quality manual
  - How the equipment has to be labeled
  - What must be arranged

Slide 28

It has to be shown by means of internal audits, that the operation (i.e. the daily work in the laboratory) is in compliance with the quality system. This must be stated in the quality manual: A schedule has to be defined and a standard procedure for internal audits has to be prescribed.

Internal Audits

- Internal Audits can show that the operation of the laboratory is in compliance with its quality system and with ISO/IEC 17025
- Schedule
- Prescribed procedure
- Schedule and procedure must be documented in the quality manual

Slide 29

Regular management reviews are required by ISO/IEC 17025. They can show that the management system and the testing activities are suitable and effective. They have to be planned. The schedule and the procedure have to be documented in the quality manual.

Management Review

- A management review can show the continuing suitability and effectiveness of the management system and of the testing activities
- Schedule
- Prescribed procedure
- Schedule and procedure must be documented in the quality manual

Slide 30

Interlaboratory tests are a powerful tool for external quality control (see chapter 15). A laboratory should take part wherever possible. The participation has to be planned. The results and if necessary the corrective actions have to be documented. Procedures have to be accordingly described in the quality manual.

Interlaboratory Tests

- Participation in interlaboratory test should be a matter of course for each laboratory in its testing field
- Planning
- Results
- Corrective actions

Slide 31

The quality manual should also contain procedures on how complaints from clients or from other parties are handled. The laboratory should have a general policy, on how to deal with complaints.

Complaints

- Policy and procedure for the handling of complaints
  - From clients
  - From other parties

Slide 32

ISO/IEC 17025 requires a system for document control for all kinds of documents in the laboratory.

Document Control - General ISO/IEC 17025 – 4.3.1

- The laboratory shall establish and maintain procedures to control all documents that form part of its management system (internally generated or from external sources), such as
  - regulations,
  - standards,
  - other normative documents,
  - test and/or calibration methods, as well as
  - drawings,
  - software,
  - specifications,
  - instructions and
  - manuals.

Slide 33

What does "document control system" mean? Every document must be clearly identifiable. This includes a unique number and revision status to show if this document has been revised. It could be useful to include the type of document into the number to ease the operation of the system. This is demonstrated on the slide with an SOP as example.

Document Control System

- Every document must be clearly identified in the control system
- Together with its revision status
- This can be done e.g. by a numbering system that includes the revision (e.g. SOP-SA-132-3.1)

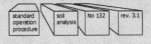

| standard operation procedure | soil analysis | No 132 | rev. 3.1 |

Slide 34

ISO/IEC 17025 requires that all documents are reviewed and approved by authorized personal prior to issue. This could be done at the bottom of the first page.
A master list of all available documents with their revision status must be established <u>and</u> be available to all personnel to avoid the use of invalid documents

Document Approval and Issue ISO/IEC 17025 – 4.3.2.1

- All documents issued to personnel in the laboratory as part of the management system shall be reviewed and approved for use by authorized personnel prior to issue.
- A master list or an equivalent document control procedure identifying the current revision status and distribution of documents in the management system shall be established and shall be readily available to preclude the use of invalid and/or obsolete documents.

Slide 35

The master list has to be included in
the quality manual or be referred to
there.

Master List

- There can be e.g. a master list of all
  valid documents together with their
  revision status in the quality manual (or
  in a separate document)

Slide 36

In this slide the requirements of
ISO/IEC 17025 for the document
approval and issue are shown. The
next slides will show details following
from these requirements.

**Document Approval and Issue
ISO/IEC 17025 – 4.3.2.2**

- *The procedure(s) adopted shall ensure that:*
  - *authorized editions of appropriate documents are
    available at all locations where operations essential to
    effective functioning of the laboratory are performed;*
  - *documents are periodically reviewed and, where
    necessary, revised to ensure that continuing suitability
    and compliance with applicable requirements;*
  - *invalid or obsolete documents are promptly removed
    from all points of issue or use, or otherwise assured
    against unintended use;*
  - *obsolete documents retained for either legal or
    knowledge preservation purposes are suitably marked.*

Slide 37

Authorized copies (that will auto-
matically be revised if the quality
manual changes) should at least be
available in the quality manager's
office, in the laboratory manager's
office and to all laboratory staff.

Availability

- Authorized copies of the quality manual
  should be available
  - In the quality manager's office
  - In the laboratory manager's office
  - In the laboratory for all staff

Slide 38

Obsolete documents have to be removed immediately and all outdated authorized copies must be marked. But they also have to be retained in the archive for knowledge preservation, sometimes also for legal reasons.

Obsolete Documents

- Have to be removed immediately
- It must be assured, that they are removed in <u>all</u> authorized copies
- Obsolete documents must be marked
- But they have to be retained for
  - Legal purposes
  - Knowledge preservation

Slide 39

ISO/IEC 17025 requires a unique identification of all management system documents with the date of issue, revision identification, page numbering, total number of pages on each page and the issuing authority.

Document Approval and Issue ISO/IEC 17025 – 4.3.2.3

- *Management system documents generated by the laboratory shall be uniquely identified.*
- *Such identification shall include*
  - *the date of issue and/or*
  - *revision identification,*
  - *page numbering,*
  - *the total number of pages or a mark to signify the end of the document, and*
  - *the issuing authority(ies).*

Slide 40

Changes of documents have to be reviewed by designated personnel.

Document Changes ISO/IEC 17025 – 4.3.3.1

- *Changes to documents shall be reviewed and approved by the same function that performed the original review unless specifically designated otherwise.*
- *The designated personnel shall have access to pertinent background information upon which to base their review and approval.*

Slide 41

It can be useful to mark altered or new text in changed documents for the reader to see that changes have been made.

> **Document Changes**
> **ISO/IEC 17025 – 4.3.3.2**
>
> - *Where practicable, the altered or new text shall be identified in the document or the appropriate attachments.*

Slide 42

Amendments of documents must be clearly marked and the revised documents shall be re-issued as soon as possible.

> **Document Changes**
> **ISO/IEC 17025 – 4.3.3.3**
>
> - *If the laboratory's documentation control system allows for the amendment of documents by hand pending the re-issue of the documents, the procedures and authorities for such amendments shall be defined.*
> - *Amendments shall be clearly marked, initialled and dated.*
> - *A revised document shall be formally re-issued as soon as practicable.*

Slide 43

If the documents are maintained in computerized systems there must be procedures to describe how this has to be done and controlled.

> **Document Changes**
> **ISO/IEC 17025 – 4.3.3.4**
>
> - *Procedures shall be established to describe how changes in documents maintained in computerized systems are made and controlled*

Slide 44

To fulfil the requirements of ISO/IEC 17025 it is useful to create a uniform header for quality documents and especially for the quality manual chapters, which e.g. could contain on the first page:

- Identification of the laboratory
- Notice that this is a part of the quality manual
- Number of the chapter
- Title of the chapter

Header of the Quality Manual

- At least on the first page of each chapter:
  - Identification of the laboratory
  - A statement that this is a part of the quality manual
  - Number of the chapter
  - Title of the chapter

Slide 45

On all pages the header should contain the date of issue, the revision number, the author's name, the reviewing person, the approval notice, the page number and the total number of pages.

Header of the Quality Manual

- On each page of each chapter:
  - Date of issue
  - Revision number
  - Name of author
  - Eventually name of person, who checked the content
  - Approval notice
  - Page number
  - Total number of pages of the chapter

Slide 46

This slide shows an example for the header of a quality manual chapter.

Example – Top and Bottom of a Quality Manual Page

Laboratory Logo	Quality Manual	chapter: 1
		revision: 1
	Quality Policy	page 1 of 3

author:	checked:	approved:	date of issue:

Slide 47

The structure of each individual quality manual is a matter for the author. But it is very useful to separate the chapters so that they can be revised individually.

> **Structure of the Quality Manual**
>
> • It is up to the author of the quality manual to decide about the detailed structure of "his" quality manual.
> • But it is extremely useful to separate it in chapters, which can be revised separately without revising and renumbering the whole manual.

Slide 48

If the quality manual is written and used in electronic form, first of all the same requirements have to be met. There must be a system for managing the access rights on the files, a system to ensure that no unauthorized documents is placed on the system, that all withdrawn documents are stored for later information and that all documents are securely backed up.

> **Quality manual in electronic form**
>
> • The same requirements have to be fulfilled
> • Special emphasis must be put on
>   • Unauthorized changes of documents
>   • Tracking of withdrawn documents
>   • Management of access rights (read/write and read only)
>   • Backup

Biliography

Grimes KR (2003) ISO 9001:2000 a Practical Quality Manual Explained. James Bennett Pty

ISO/IEC 17025:2005 – General requirements for the competence of testing and calibration laboratories

ISO 9001:2000 – Quality management systems – Requirements

Scheutwinkel M., Kindler M (2001) Quality Manual ISO/IEC 17025, MediVision, Berlin

8 Basic Statistics

Michael Koch

Slide 1

Statistics are not "l'art pour l'art". Statistics are a tool to answer questions. They can be a very useful tool, especially for analysts. Examples of these questions are: How accurate is the analysis? How many analyses do I have to make to overcome problems of inhomogeneity or imprecision? Does the product meet the specification? What is the confidence that the limit is exceeded? And many more questions in the daily work of chemical analysts can be answered using statistics.

Statistics

- Are a tool to answer questions:
  - How accurate is the analyses?
  - How many analyses do I have to make to overcome problems of inhomogeneity or imprecision?
  - Does the product meet the specification?
  - With what confidence is the limit exceeded?

Slide 2

As all analysts know, there are always variations in analytical results. These are due to unavoidable variations during the measurement, variations in the readings of the instruments as well as variations in the handling of the samples by the analyst. But on the other hand there are also variations due to the inhomogeneity of the sample that has to be analysed. The sub-sampling necessary for the analysing process results in different sub-samples. It is very important to differentiate between these two possibilities, because the measures that can be taken to minimize these variations can be totally different.

Analytical Results Vary

- Because of unavoidable deviation during measurement
- Because of inhomogeneity between the subsamples

- It is very important to differentiate between these two cases
- Because the measures to be taken can be different

Slide 3

These variations are always present. If there seems to be no variation, the resolution is not high enough. In the first column the presentation of the result is limited to 3 digits and the results are all identical. If one digit is added, some variation can be recognised showing that the results vary between 321.5 and 322.4. With one more digit nearly all results are different.

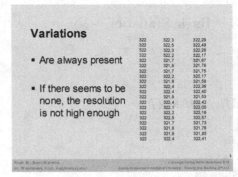

Slide 4

A histogram is used to describe the variability of results. For this purpose the range of possible values for the results is divided in a number of groups. Each group should have the same range. When the measurements are complete, count the number of observations in each group and draw a box-plot of the frequencies vs. the groups.

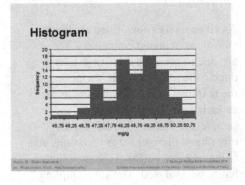

Slide 5

This diagram shows a histogram of measurements with variations between 45.5 and 51 mg/g of a specific analyte. Most of the measurements are between 48 and 50 mg/g. It can be seen, that it is necessary to have a large number of measurements to be able to draw a histogram.

Slide 6

If the number of measurements is even
larger and the range of values covered
by each group is small enough, the
histogram changes into a distribution
curve. In practice, a distribution curve
is a mathematical model for the
measurement data.

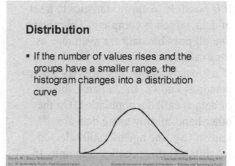

Slide 7

Summing up all the frequencies gives
the cumulative distribution. The cumulative distribution curve is the integral
of the distribution curve. This curve
can be used to find the proportion of
the values that fall above or below a
certain value. For distribution curves
or histograms, which have a maximum,
(typical for all analytical measurements), the cumulative distribution
curves show the typical S-shape.

Slide 8

The cumulative distribution curve of
the above example shows that
approximately 50% of the values are
below 48.5 and 50 % are above this
value. And for example it can also be
seen that only about 10 % of the
values are below 46,5 mg/g. Also the
above mentioned typical S-shape can
be seen.

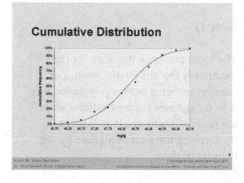

Slide 9

It is possible to apply statistics to a set of data, which is complete. This means that all possible data are available, as for example, the age of all inhabitants of a town, where the age of every single person is known. In statistics, this set of data is called a population. On the other hand, if only some data are known (e.g. the age of 1000 selected inhabitants) this is called a sample. If the whole population is known, parameters like the mean can be calculated. If only a sample is known, these parameters have to be estimated. For analytical data, it is only possible to know a sample of all possible measurement results, because the number of these is infinite.

Population – Sample

- It is important to know, whether all data (the population) or only a subset (a sample) are known.

Sample	Population
A selection of 1000 inhabitants of a town	All inhabitants of a town
Any number of measurements of copper in a soil	*Not possible*

Slide 10

One major task of statistics is to describe the distribution of a set of data. The most important characteristics of a distribution are the location, the dispersion, the skewness and the kurtosis. These are discussed in the following slides.

Distribution Descriptives

- The data distribution can be described with four characteristics:
  - Measures of location
  - Measures of dispersion
  - Skewness
  - Kurtosis

Slide 11

The most common measure for the location is the arithmetic mean. For random samples from a population this is in most cases the best estimation of the population mean. It is calculated by dividing the sum of all data by the number of data.

Location – Arithmetic Mean

- Best estimation of the population mean μ for random samples from a population

$$\bar{x} = \frac{\sum_{i=1}^{n} x_i}{n}$$

Slide 12

For the calculation of the median the data have to be sorted by their value. The median is the central value of this series. If the number of data is even, the median is the arithmetic mean of the central values. The median is a so-called robust parameter, because extreme values or outliers do not affect it.

Location –
Median

- Sort data by value
- Median is the central value of this series

- The median is robust, i.e. it is not affected by extreme values or outliers

Slide 13

The mode is the most frequent value. To calculate the mode it is necessary to have a large number of data with at least partially the same value. Because it is possible to have several data sets with the same value at the same number, it is possible to have more than one mode in the data distribution.

Location –
Mode

- The most frequent value

- Large number of values necessary
- It is possible to have two or more modes

Slide 14

Besides the above-mentioned measures for the location there are several other possible parameters. The geometric mean is the arithmetic mean of the logarithms of the data (or the nth root of the product of all data). The harmonic mean is the reciprocal of the mean of the reciprocals of all single values.

Methods for robust statistics have been developed that deliver good results (i.e. estimation of the population mean) even with a relatively large number of outliers or with a skewed distribution. For more detailed descriptions of these methods please refer to the relevant textbooks.

Location –
Other Means

- Geometric mean
- Harmonic Mean
- Robust Mean (e.g. Huber mean)
- …

Slide 15

The most important measure for the dispersion of a data distribution is the variance. The variance of a population (with all possible data being known) is the mean of the squares of the deviations of the individual values from the population mean.

Dispersion –
Variance of a Population

- The mean of the squares of the deviations of the individual values from the population mean

$$\sigma^2 = \frac{\sum_{i=1}^{n}(x_i - \mu)^2}{n}$$

Slide 16

If not all possible data are known and hence the population mean and the dispersion have to be estimated from a sample, the variance is calculated in a very similar way. In the formula for the variance of a population, the number of data n is replaced by (n-1) and the population mean μ by the arithmetic mean $\bar{x}$.

Dispersion –
Variance of a Sample

- Variance of population with (n-1) replacing n and the arithmetic mean of the data (estimated mean) replacing the population mean μ

$$s^2 = \frac{\sum_{i=1}^{n}(x_i - \bar{x})^2}{n-1}$$

Slide 17

The standard deviation, for the population as well as that estimated from a sample, is the positive square root of the variance.

Dispersion –
Standard Deviation

- The positive square root of the variance

Population   Sample

$$\sigma = \sqrt{\frac{\sum_{i=1}^{n}(x_i - \mu)^2}{n}} \qquad s = \sqrt{\frac{\sum_{i=1}^{n}(x_i - \bar{x})^2}{n-1}}$$

Slide 18

If means are calculated from samples of the data, they are more closely clustered than the data themselves, because the scattering of the data is eliminated to some extent by calculating the means. The standard deviation of the mean therefore is the standard deviation divided by the square root of the number of measurements used to calculate the mean.

**Dispersion –
Standard Deviation of the Mean**

- Possible means of a data set are more closely clustered than the data

$$sdm = \frac{\sigma}{\sqrt{n}}$$

where n is the number of measurements for the mean

Slide 19

In many cases the standard deviation in relation to the mean is more interesting than the absolute value. The relative standard deviation RSD therefore, is the comparison of the standard deviation to the mean and it is often calculated in percent.

**Dispersion –
Relative Standard Deviation**

- measure of the spread of data compared with the mean

$$RSD = \frac{s}{\bar{x}}$$

Slide 20

Randomly selected data from a population are normally symmetrically distributed around the mean value. But in some cases the distribution is unsymmetrical; the distribution is skewed. Data from chemical analyses close to the limit of detection are unavoidably skewed, because negative concentrations are not possible.

Skewness

- The skewness of a distribution represents the degree of symmetry
  - Analytical data close to the detection limit often are skewed because negative concentrations are not possible

Slide 21

There is a fourth parameter to describe the characteristics of a distribution, which is the kurtosis. This is the peakedness of the data set. If the data set has a flat peak in the distribution curve it is called platykurtic. If the peak is very sharp it is called leptokurtic. Distributions with peaks in-between are called mesokurtic.

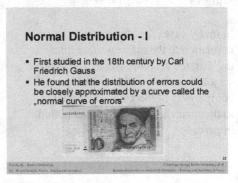

Slide 22

The normal distribution and the statistical tools linked with it are the most important statistical tools in analytical chemistry. The normal distribution was first studied by the German mathematician Carl Friedrich Gauss as a curve for the distributions of errors. In putting his picture, his formula and the curve on the former 10 DM bill the German government recognized the importance of his studies.

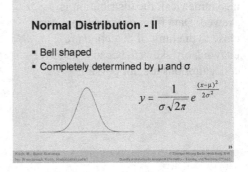

The discussion of other distributions that are important in some circumstances goes beyond the scope of this book and the reader is referred to statistics textbooks to find more details.

Slide 23

The curve of the normal distribution is bell shaped and is completely determined by only two parameters, the central value μ and the standard deviation σ.

Normal Distribution - II

- Bell shaped
- Completely determined by μ and σ

$$y = \frac{1}{\sigma\sqrt{2\pi}} e^{\frac{(x-\mu)^2}{2\sigma^2}}$$

Slide 24

The normal distribution has some properties that are very important in understanding statistical results. The curve is symmetrical about the central value μ. 68,27% of the values lie within μ ± 1σ, 95,45% within μ ± 2σ and 99,73% within μ ± 3σ.

Normal Distribution – Important Properties

- The curve is symmetrical about μ
- The greater the value of σ the greater the spread of the curve
- Approximately 68% (68,27%) of the data lie within μ±1σ
- Approximately 95 % (95,45%) of the data lie within μ±2σ
- Approximately 99,7 % (99,73%) of the data lie within μ±3σ

$$y = \frac{1}{\sigma\sqrt{2\pi}} e^{\frac{(x-\mu)^2}{2\sigma^2}}$$

Slide 25

For many purposes in analytical chemistry the normal distribution is a very useful model. The area under the normal curve can be divided in various sections characterized by the arithmetic mean and a multiple of the standard deviation. These areas can then be interpreted as proportions of observations falling within ranges defined by specific probabilities.

Normal Distribution – a Useful Model

- With the mean and the standard deviation area under the curve can be defined
- These areas can be interpreted as proportions of observations falling within these ranges defined by μ and σ

Slide 26

If one has a specific problem, which can be converted into a statistical question, it is important to know if the question is one-tailed or two-tailed. Typically one-tailed are questions for limits, such as: "Is the limit exceeded or not?" (e.g. alcohol in blood). If we want to know whether a measured value lies within a certain range this is a typically two-tailed question. The value could be too low or too high.

One-tailed / Two-tailed Questions

- One-tailed
  - E.g. a limit for the specification of a product
  - It is only of interest, whether a certain limit is exceeded or not
- Two-tailed
  - E.g. the analysis of a reference material
  - We want to know, whether a measured value lies within or outside the certified value ± a certain range

Slide 27

In this figure the one-sided probabilities are shown for a couple of multiples of σ. The probability that a value is lower than μ + σ is 84.1%. Or, the other way round: With a probability of 95%, the value is lower than μ + 1,64σ and with a probability of 5% it is higher than μ + 1,64σ. Please note: The low probability that a value may be e.g. below μ - 3σ does not matter in this case. The only question is: What is the probability that the value exceeds μ + kσ?

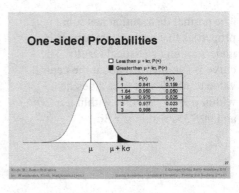

Slide 28

For two-sided probabilities we get the well-known values of a probability of 68.3% for a value being within the range of μ ± σ, 95.4% for the range μ ± 2σ and 99.7% for the range μ ± 3σ. In analytical chemistry the 90 and 95% values very often are used. These correspond to k-values of 1.64 and 1.96 respectively. Note: If a value is given together with its standard deviation (or in most cases an estimation of this value), the probability for a single measurement lying within ± one standard deviation is only 68.3%.

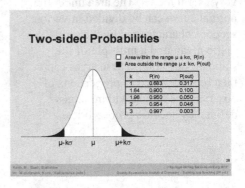

Slide 29

It may be very difficult to see from small random samples if a population is normally distributed. In this slide an example is given with samples from a population of 996 data. The population mean is 100 and the standard deviation is 19.75. First, 10 data are randomly sampled (about 1% of the data) resulting in an estimation for the mean of

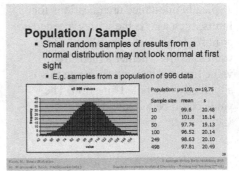

99.6 and an estimated standard deviation of 20.48. This sample does not look as though it is normally distributed. Even when we randomly take 50% of the values, the estimated values for the mean and the standard deviation are not much closer to the true value although the data does look well distributed.

Slide 30

The result of a chemical analysis is never a single value, but a range in which the true value is supposed to lie with a given confidence. This range is called confidence interval or confidence limits. The greater the confidence required the larger the interval will be.

Confidence Limits / Intervals

- The confidence limits describe the range within which we expect with given confidence the true value to lie.

Slide 31

If the population standard deviation is known and the mean is estimated from one measurement, for a confidence level of 95 % the true value will lie in the range of $\overline{x} \pm 1.96 \cdot \sigma$. If the mean is estimated from n measurements the range has to be divided by $\sqrt{n}$.

Confidence Limits for the Population Mean

- If the *population* standard deviation σ and the estimated mean are known:
- CL for the population mean, estimated from n measurements on a confidence level of 95%

$$CL = \overline{x} \pm 1.96 \cdot \frac{\sigma}{\sqrt{n}}$$

Slide 32

Normally the population standard deviation σ is not known, and has to be estimated from a sample standard deviation s. This will add an additional uncertainty and therefore will enlarge the confidence interval. This is reflected by using the Student-t-distribution instead of the normal dis-tribution. The t value in the formula can be found in tables for the required confidence limit and n-1 degrees of freedom.

Confidence Limits for the Population Mean

- If the standard deviation has to be estimated from a sample standard deviation s (this is normally the case):
- Additional uncertainty → use of Student t distribution

$$CL = \overline{x} \pm t \cdot \frac{s}{\sqrt{n}}$$

Where t is the two-tailed tabulated value for (n-1) degrees of freedom and the requested confidence limit

Slide 33

The table on this slide is an example for such t-values. In the first two rows are the confidence levels for one-tailed (1T) and two-tailed (2T) questions. The first column shows the degrees of freedom, which is obtained from the number of measurements. Tables with t-values might be found in all statistics textbooks and relevant standards.

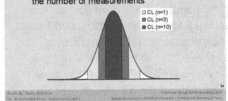

Slide 34

As can be seen in the formula of the previous slide the confidence limit depends very much on the number of measurements. This is demonstrated in this slide showing the confidence limits for a single measurement, for three and for ten measurements

**Confidence Limits –
Depending on n**

- Confidence Limits are strongly dependent on the number of measurements

Slide 35

When talking about quality of chemical measurements trueness, precision, accuracy and error are some of the more important keywords. Therefore a clear definition is necessary (see also chapter 11, slides 36 and 44).

Trueness – Precision – Accuracy - Error
- Trueness
  - The closeness of agreement between the true value and the sample mean
- Precision
  - The degree of agreement between the results of repeated measurements
- Accuracy
  - The sum of both, trueness and precision
- Error
  - The difference between one measurement and the true value

Slide 36

The linkage between these keywords is shown in this figure with a target as example.

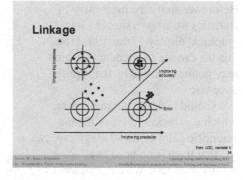

Slide 37

Significance testing is another topic of statistics. What does this mean? A typical situation for analytical chemists is where a reference material together with its certified value and uncertainty is given on the one hand and on the other hand there is the result of the analysis. The analyst wants to know whether the bias of his result is significant or just due to random errors.

Significance Testing - ?

- Typical Situation:
  - We have analysed a reference material and we want to know, if the bias between the mean of measurements and the certified value of a reference material is „significant" or just due to random errors.
- This is a typical task for significance testing

Slide 38

The aim of significance testing is to make a decision about a population that is based on observations made from a sample from this population. This decision can only be made with a certain level of confidence. Therefore there can never be a guarantee of the correctness of the decision. If there is 95% confidence, there is 5% doubt.

Significance Testing

- A decision is made
  - About a population
  - Based on observations made from a sample of the population
  - On a certain level of confidence
- **There can never be any guarantee that the decision is absolutely correct**
- On a 95% confidence level there is a probability of 5% that the decision is wrong

Slide 39

Before we apply the machinery of statistics we always should inspect a graphical display of the data. From this we can see many details, which may not be revealed by the significance test:
We should look for "strange" data, which e.g. make no sense from a scientific point of view. Maybe we can see that there is no need at all for a significance test. We can crosscheck the statistical result with our common sense visual appraisal. And the most essential point: we possibly can see, that there is statistical significance but the difference is not important in magnitude compared with our quality needs.

Before Significance Testing

- Inspect a graphical display of the data before using any statistics
- Are there any data that are not suitable?
- Is a significance test really necessary?
- Compare the statistical result with a commonsense visual appraisal
- A difference may be statistically significant, but not of any importance, because it is small

Slide 40

The example in the last slide shows one of the fundamental points in the application of statistics to analytical results. Statistics are very useful tools to solve a lot of problems. But nevertheless they should always be applied with a critical view of the results, whether they make scientific sense.

The most important thing in statistics

No blind use

Slide 41

Significance testing can be divided into a small number of steps. It starts with the formulation of the Null hypothesis. This is the assumption, which is made about the properties of a population of data expressed mathematically, e.g. "there is no bias in our measurements". The second step is the formulation of the alternative hypothesis, the opposite of the Null hypothesis, in the above example: "there is a bias".

The Stages of Significance Testing - I

- Null hypothesis
  - A statement about the population, e.g.: there is no bias ($\mu = x_{true}$)

- Alternative hypothesis H_1
  - The opposite of the null hypothesis, e.g.: there is a bias ($\mu \neq x_{true}$)

Slide 42

In the next step a value for the test is calculated from the data and compared with the tabulated critical value. If the calculated value exceeds the critical value this indicates significance. To finalize the significance test the test statistics have to be evaluated with respect to the Null hypothesis. This enables us to make decisions and to draw conclusions.

The Stages of Significance Testing - II

- Critical value
  - The tabulated value of the test statistics, generally at the 95% confidence level
  - If the calculated value is greater than the tabulated value than the test result indicates a significant difference
- Evaluation of the test statistics
- Decisions and conclusions

Slide 43

In the next few slides an example for significance testing in analytical chemistry is shown. The first point is the question that we want to ask. Is the result of our measurements (i.e. the mean) significantly different from the certified value of a reference material or is the difference just due to random variations?

Significance Test Example – the Question

- Is the bias between the mean of measurements and the certified value of a reference material significant or just due to random errors?

Slide 44

Before starting the calculations we can examine a graphical display of the results.
The results of the measurements are shown as dots on the scale, the mean is marked as $\bar{x}$. The vertical lines show the 95% confidence limits. In the upper case the certified value is outside these limits, therefore the bias is significant. In the lower case the certified value is inside the confidence interval, the bias is insignificant.

Significance Test example – Graphical Explanation

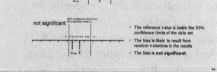

Slide 45

Let us translate this into mathematics. The Null hypothesis (our assumption) says: There is no bias. We calculate the confidence limits as shown above and translate our Null hypothesis into a mathematical formula resulting in a formula for an observed Student-t-factor $t_{observed}$. We can now compare this observed value with the critical value for 95% confidence and the degrees of freedom for our number

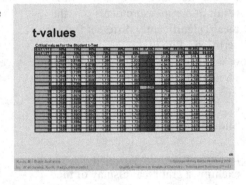

of measurements. Since the reference values of a CRM only is a estimate of the true value, we have also have to consider the uncertainty of this value. This is done by extending the denominator in the last equation.

Slide 46

We get the critical value from the table of t-values. We have a two-tailed question, we want to have 95% confidence level, and with 10 measurements we have 9 degrees of freedom. Therefore we get from the table a critical value of 2.262.

t-values

Slide 47

Now we can draw the conclusions: If the critical value $t_{critical}$ exceeds the observed value $t_{observed}$ we can say: there is no significant bias at the 95% level or one of the other formulations in the slide. But we cannot say: there is no bias.

Significance Test Example – Conclusions I

- If we find $t_{critical} > t_{observed}$
  - Conforming to our Null hypothesis
- We can say:
  - We cannot reject the Null hypothesis
  - We accept the Null hypothesis
  - We find no significant bias at the 95% level
  - We find no measurable bias (under the given experimental conditions)
  - But not: there is no bias

Slide 48

On the other hand, if we find $t_{critical}$ smaller than $t_{observed}$ we can say: "we find significant bias at the 95% level" or "we reject the null hypothesis". But we cannot say: there is bias. Since our calculations are based on a 95% confidence level, there is a chance of about 5% that we are wrong.

Significance Test Example – Conclusions II

- If we find $t_{critical} < t_{observed}$
  - Not conforming to our Null hypothesis
- We can say:
  - We reject the null hypothesis
  - We find significant bias at the 95% level
  - **But not: there is a bias,** because there is roughly a 5% chance of rejecting the Null hypothesis when it is true.

Slide 49

The example we just looked at is called one-sample t-test. It compares the mean of analytical results with a stated value. This is a typical analytical question. The problem may be two-tailed as in our example, where it doesn't matter, if the analytical value is biased to the one or the other direction. Or the question could be one-tailed, e.g. if we want to know whether the copper content analysed in an alloy is below the specification.

One-sample t-test

- This test compares the mean of analytical results with a stated value
- The question may be two-tailed
  - As in the previous example
- Or one-tailed
  - E.g.: is the copper content of an alloy below the specification?
  - For one-tailed questions other t-values are used

Slide 50

If we have a one-tailed question we have to use other t-values. In the table shown in this slide we have to select the level of confidence from the first row for one-tailed and from the second row for two-tailed questions.

t-values

Slide 51

If we don't have one stated value, but two independent sets of data (e.g. two analytical results from different laboratories or methods) we have to use the two-sample t-test, because we have to consider the dispersion of both data sets. In the same way as above we have to look carefully, what our question is: it may be two-tailed (are the results significantly different?) or one-tailed (is the result from method A significantly lower than that from method B?)

Two-sample t-test

- This test compares two independent sets of data (analytical results)
- The question may be two-tailed or one-tailed as well
  - Two-tailed: are the results of two methods significantly different
  - One-tailed: are the results of method A significantly lower than the results of method B

Slide 52

The mathematics for the two-sample t-test looks somewhat more complex. Beneath the difference between the two means the standard deviations of both data sets and the number of data are involved in the formula. The degree of freedom is given by n_1+n_2-2.

Two-sample t-test – t-value

- The formula for the calculation of the t-value is somewhat more complex:

$$t_{observed} = \frac{(x_1 - x_2)}{\sqrt{\left(\frac{1}{n_1} + \frac{1}{n_2}\right) \times \left(\frac{s_1^2(n_1-1) + s_2^2(n_2-1)}{(n_1+n_2-2)}\right)}}$$

- The degree of freedom for the tabulated value $t_{critical}$ is
  d.o.f = n_1+n_2-2

Slide 53

But there is a prerequisite that has to be fulfilled, before we can apply the two-sample t-test. It is valid only, if the standard deviations of the data sets are not significantly different. This has to be tested using the F-test.

Two-sample t-test - Validity

- The two-sample t-test is valid only if there is no significant difference between the both standard deviations s_1 and s_2
- This can (and should) be tested with the F-test

Slide 54

The F-value is calculated from the estimated variances s^2 of the two data sets. The critical value $F_{critical}$ can also be found from a table. For that we need the degrees of freedom for both data sets.

F-test

- The F-test is a method of testing whether two independent estimates of variance are significantly different
- The F-value is calculated according to $F_{observed}=s_A^2/s_B^2$, with $s_A^2 > s_B^2$
- The critical value of F depends on the degrees of freedom $dof_A=n_A-1$ and $dof_B=n_B-1$ and can be taken from a table

Slide 55

In the example shown in this slide we have an F-table for 95% confidence. We assume that there are 10 measurements for each data set. The degrees of freedom are 9 for both. Therefore we find the critical F-value to be 4.03.

F-values

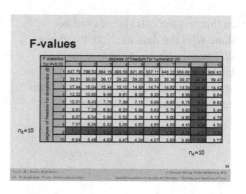

Slide 56

From the comparison of our observed F-value and the critical F-value we can make our decision. If the observed value is smaller than the critical value we can say: The estimated variances are not significantly different at the chosen confidence level.

F-test - Decision

- If $F_{observed} < F_{critical}$ than the variances s_A^2 and s_B^2 are not significantly different
- On the chosen confidence level

Slide 57

The two-sample t-test is only applicable, if the data are collected from analysing the same sample several times. If the data come from the analysis of different samples where the differences between the samples are much greater than the differences between the two methods or the two laboratories, it is necessary to apply the paired t-test.

Slide 58

In this test the calculations are not done with the data itself, but with the differences between the two methods/ laboratories. These differences are tested for a significant difference from zero with a one-sample t-test.

Slide 59

As described in slide 45 the value $t_{observed}$ is calculated with $x_{ref} = 0$ and the critical value $t_{critical}$ is taken from a table as shown above.

Slide 60

In summary it can be stated, that there are many helpful statistical tools for the analytical chemist. Although these tools cannot be a "cure-all", every analyst should know how to handle these tools in order to answer his and his customers' questions.

Statistics

- There are many statistical tools for the analyst to interpret analytical data
- Statistics are not a "cure-all" that can answer all questions
- If the analyst uses statistics with a commonsense, this is a powerful tool to help him to answer his customers or his own questions

Slide 61

There is a lot of literature about statistics and especially for analytical chemists. As an example two useful books and training software are recommended here for further details about statistical tools for analytical chemists.

Recommended literature and training software

- LGC, VAMSTAT II, Statistics Training for Valid Analytical Measurements. CD-ROM, Laboratory of the Government Chemist, Teddington 2000
- H. Lohninger: Teach/Me - Data Analysis. Multimedia teachware. Springer-Verlag 1999
- J.N. Miller and J.C. Miller: Statistics and Chemometrics for Analytical Chemistry. Prentice Hall Publisher 2005
- T.J. Farrant: Practical Statistics for the Analytical Scientist – a Bench Guide. Royal Society of Chemistry for the Laboratory of the Government Chemist (LGC), Teddington 1997

Bibliography

Farrant TJ (1997) Practical Statistics for the Analytical Scientist – a Bench Guide. Royal Society of Chemistry for the Laboratory of the Government Chemist (LGC), Teddington

LGC (2000) VAMSTAT II – Statistics Training for Valid Analytical Measurements. CD-ROM, Laboratory of the Government Chemist, Teddington 2000

Lohninger H (1999) Teach/Me - Data Analysis. Multimedia teachware. Springer-Verlag

Miller JN, Miller JC (2005) Statistics and Chemometrics for Analytical Chemistry. Prentice Hall Publisher

9   Calibration

Michael Koch

Slide 1

Most chemical analytical methods
need a calibration to convert the
meaurement signal into an analytical
result. This chapter describes the
basics of calibration, different
possibilities, how to perform it and the
information derived from it.
The detection and quantification
capabilities of analytical methods
often are important if they are used at
trace levels of analytes. The description

Contents

- Introduction
- Basics of Calibration
- Limits of detection, quantification
- Standard Addition Method

of the standard addition method, a special calibration in the sample finalises the
chapter.

Slide 2

Calibration is the link between the
signal of a measurement (e.g. extinction
in photometry or peak area from a
detector in chromatography) and the
quantity we intend to measure.
If we use a standard solution traceable
to a stated reference, we ensure the
traceability of our measurement by
extending the traceability chain (for
more details on traceability see chapter
10).

Introduction

- Calibration is an important process in
  - Establishing the link between a signal of
    the measuring instrument and the
    associated quantity (e.g. concentration) of
    the measurand
  - Establishing traceability
  - Method validation to get the performance
    characteristics

From the basic calibration of our method we can derive some performance characteristics of the method. This is important for method validation (see chapter 11)

B.W. Wenclawiak et al. (eds.), *Quality Assurance in Analytical Chemistry: Training
and Teaching*, DOI 10.1007/978-3-642-13609-2_9, © Springer-Verlag Berlin Heidelberg 2010

Slide 3

Analytical measurements are based on the fact that the response actually being measured varies with what we want to determine.
Measurement generally means comparison of the unknown with the known. So, in calibration we compare the signal from the unknown amount of analyte in our sample with the signal of the known amount in our calibration sample.

What is Calibration?

- Calibration is the process of establishing how the response of a measurement process varies with respect to the parameter being measured
- The usual way to perform calibration is to subject known amounts of the parameter (e.g. using a measurement standard or reference material) to the measurement process and monitor the measurement response

Slide 4

Of course we want to cover a certain range of the quantity we want to measure (e.g. concentration) with our measurement (the working range). So we need a mathematical function between the signal and the concentration range valid over the whole working range. But we also get statistical information and other characteristics from the calibration data.

Two Major Aims

- Establishing a mathematical function which describes the dependency of the system's parameter (e.g. concentration) on the measured value
- Gaining statistical information and characteristics of the analytical system, e.g. sensitivity, precision

Slide 5

There are different concepts for the calibration. External calibration uses the measurement of separate samples containing known amounts of analyte and we compare the signal from both, the calibration sample and our unknown sample. This method requires that the differences in the matrix of sample and calibration sample does not significantly influence the response. The use of an internal

Calibration Concepts

- External standard
- Internal standard
- Standard addition

standards includes addition of a known amount of a substance, which has properties similar to the analyte. Quite often internal standards are used to correct for
insufficient recovery rates.

The standard addition procedure is a calibration in the sample. Known amounts of
analyte are added to the sample and from the increase in signal the original content
is extrapolated (see slide 47)

Slide 6

With calibration we define the working range, where we are sure that we
can analyse the sought component in a
way which is fit for the intended
purpose.

Introduction

Goals of Calibration

- "Ability to calculate a (measurement)
  result in a secure (safe) working range"

(Funk, W., Dammann, V., and Donnevert, G.: "Quality
Assurance in Analytical Chemistry")

Slide 7

This slide describes the steps we have
to follow. We start with choosing the
range we want to cover with our
analysis. Than we measure some
calibration standards in this range. We
try to apply a linear regression, but we
also check for the validity of this
choice. If we succeed, we calculate the
performance characteristics of our
method and finally fix the working
range.

Introduction

First Steps to the Goal

- Establishing the calibration function
  - Choosing the preliminary working range
  - Measuring several calibration standards
  - Linear regression
  - Test of non linear regression
  - Test of variance homogeneity
  - Calculate performance characteristics
  - Fix the working range

Slide 8

In the routine use of the method we
calculate the measurement results from
the signal using the analytical function
derived from the calibration function.
And finally of course we report our
result.

Introduction

In Routine

- Calculating the (measurement) results
  - Conversion of the calibration function into
    an analytical function
  - Reporting the measurement results

Slide 9

The mathematical description of the link between signal and the quantity intended to be measured should be as simple as possible. If we have a linear relationship and no blank response, i.e. no signal, if the analyte is absent, we could use a simple linear function without intercept. If we cannot exclude that we have a blank signal we add an intercept. The calibration function is then described by slope and intercept.

Basics of Calibration

- Mathematical Functions
  - simple linear function without intercept
    $y = m x$
- linear function
  (Intercept b, slope a)
  $y = a + bx$
- quadratic function
  (intercept b, slope 2cx+b)
  $y = a + bx + cx^2$

In cases where we don't have a strict linear relationship we could use a quadratic function. These are the functions commonly used in analytical chemistry. Of course, in special cases more sophisticated functions might also be used.

Slide 10

Linear regression, or more exact, linear least-squares regression, is based on minimizing the sum of the squared vertical deviations (residuals) from the regression line.

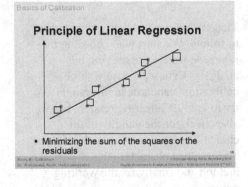

Basics of Calibration

Principle of Linear Regression

- Minimizing the sum of the squares of the residuals

Slide 11

ISO 8466 describes how to perform calibration. Part 1 is covering the linear regression and part 2 the second order calibration strategy.

Basics of Calibration

International Standards

- ISO 8466 Water quality– Calibration and evaluation of analytical methods and estimation of performance characteristics
  - Part 1: Statistical evaluation of the linear calibration function
  - Part 2: Calibration strategy for non-linear second order calibration functions

Slide 12

The calibration function which'ever chosen is describing how the signal depends on the analyte content x. The inverted function, the analytical function is used in the routine application of the method to calculate a content from the signal of an unknown sample.

Basics of Calibration

Analytical Function / Calibration Function

- With calibration standards we get a measurement result y for each content x of the analyte in the standard
  With that we get (by regression analysis) the **calibration function**

  $$y = f(x)$$

- If we invert the function we can calculate the content of analyte in the unknown sample from the result of the measurement.
  This function is the **analytical function**

  $$\hat{x} = f(\hat{y})$$

Slide 13

The basic calibration of a method only covers the final measurement step without any preceding sample preparation. Pure analytical standard solutions are used here. Of course this does not cover the whole analytical process. So method characteristics for the basic calibration are not transferable to the whole analytical process. During validation the influence of other matrix constituents has to be investigated.

Basics of Calibration

Basic Calibration

- With the Basic Calibration only the measurement step itself is calibrated
- I.e. no sample preparation like extraction, digestion etc. is done
- We simply analyse standards in pure solvent

Slide 14

Before we start we have to decide about a preliminary working range. The most important point is the purpose of the future analyses. In which range do we expect the quantities to be measured? Of course there are technical limitations. For very low concentrations we have to ensure that the signal triggered by this content is significantly different from the signal of the blank. There also might be requirements on the uncertainty of the measurement, which might not be fulfilled at the lower end of the preliminary working range. If we want to use a linear calibration strategy we have to ensure that the variances are sufficiently homogeneous and that there is no significant deviation from linearity.

Basics of Calibration

Choosing the Preliminary Working Range

- Take into consideration
  - The practical application of the analysis (the purpose)
  - The possibilities that are technically feasible
    - Measurement results at the lower application limit must be significantly different from blanks
    - The requested analytical precision (or measurement uncertainty) has to be achieved over the whole working range
    - If a linear regression procedure has to be applied the variances have to be homogeneous over the whole range and linearity has to be assured

Slide 15

The first step in the laboratory is the preparation of the standard solutions. For the basic calibration we will use sufficiently pure material which is free from interfering compounds. We have to ensure the homogeneity of the samples and the representativeness regarding the characteristics described in the slide. Of course the samples have to be sufficiently stable for the time period within which they are intended to be used. If necessary, we have to preserve the samples. Storage of the samples has to be organised in such a way that the integrity of the sample is not influenced.

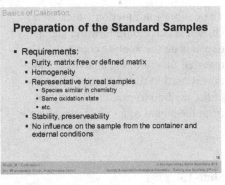

Slide 16

Preparation of standard samples always should be as accurate as possible. Therefore gravimetric procedures should be preferred compared to volumetric ones and several dilutions should be avoided since each dilution step adds to the uncertainty. For a basic calibration we need 6 to 10 standard samples. They should be distributed equidistant over the whole working range. Linear regression requires equidistant distribution. Otherwise a weighted regression would be needed.

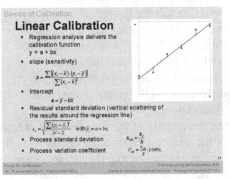

Slide 17

Now we apply the usual linear regression procedure, which delivers slope and intercept of the calibration function. From the residuals, i.e. the vertical distances of the calibration points from the regression line, the residual standard deviation can be calculated. This standard deviation is a quality indicator for the calibration function.

The correlation coefficient r^2 does *not* indicate the quality. If we divide the residual standard deviation by the sensitivity (the slope of the function) we get the process standard deviation and the process variation coefficient respectively.

Slide 18

If the relationship between the signal and the concentration is not linear, we may apply a second order calibration function. For details about this slightly more difficult calculation see ISO 8466-2.

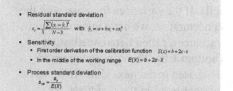

Basics of Calibration

Second Order Calibration Function
$$y = a + bx + cx^2$$

- Calculations are somewhat more difficult here
- For details see ISO 8466-2

Slide 19

The formula for the residual standard deviation is a bit different. This reflects that the degrees of freedom are one less. The slope of the function and therefore also the sensitivity is concentration dependent. So usually the sensitivity in the middle of the working range is reported. We may also calculate the process standard deviation and process variation coefficient in the middle of the working range.

Basics of Calibration

Second Order Calibration Function

- Residual standard deviation
  $$s_y = \sqrt{\frac{\sum(y_i - \hat{y}_i)^2}{N-3}} \quad \text{with } \hat{y}_i = a + bx_i + cx_i^2$$
- Sensitivity
  - First order derivation of the calibration function $S(x) = b + 2c \cdot x$
  - In the middle of the working range $E(\bar{x}) = b + 2c \cdot \bar{x}$
- Process standard deviation
  $$s_{x0} = \frac{s_y}{E(\bar{x})}$$
- Process variation coefficient
  $$V_{x0} = \frac{s_{x0}}{\bar{x}} \cdot 100\%$$

Slide 20

Usually linear calibration functions are preferred. To check for linearity the easiest thing is a visual check by inspection of the calibration data together with the regression line. Sometimes non-linearity is so obvious, that no statistical test is required.

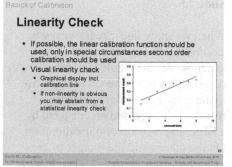

Basics of Calibration

Linearity Check

- If possible, the linear calibration function should be used, only in special circumstances second order calibration should be used
- Visual linearity check
  - Graphical display incl. calibration line
  - If non-linearity is obvious you may abstain from a statistical linearity check

Slide 21

If it is not clear, the Mandel-test may be applied for linearity check. We calculate the linear and the 2nd order calibration function and the respective residual standard deviations. If the F-test (as described in the slide) delivers a significant difference between the residual standard deviations, this shows that the 2nd order calibration function significantly better describes the calibration. So this function should be preferred. If it is not significantly better, we should use the linear function.

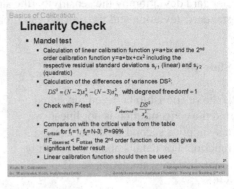

Slide 22

A graphical display of the residuals tells us a lot about our data. They should be normally distributed (top left). If the variances increase with the concentration, we have inhomogeneous variances, called heteroscedasticity (bottom left). The consequences are discussed in the next slide. If we have a linear trend in the residuals, we probably used the wrong approach or we have a calculation error in our procedure (top right). Non-linearity of data deliver the situation described on bottom right, if we nevertheless use the linear function.

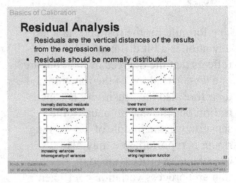

Slide 23

When we apply linear regression we assume that the variances are homogeneous over the whole concentration range. If this is not the case (as shown in the graph) , we might get wrong results. This is caused on one hand by a larger imprecision of our calibration. On the other hand the danger of getting a wrong slope is higher, leading to biased measurements at the end.

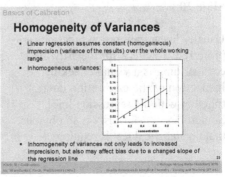

Slide 24

There is a statistical test to check the
homogeneity of variances. We repeatedly measure the highest and the
lowest standard samples (10 times
each) and calculate the variances for
both data sets. The F-test gives us an
answer on the question, whether they
are significantly different or not.
In cases where we have inhomogeneous variances we may reduce the
working range or use a regression
procedure weighted with variances at different concentrations.

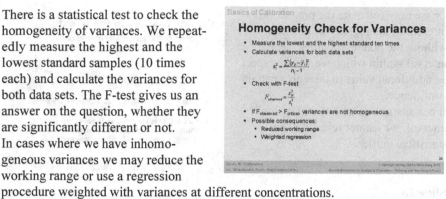

Slide 25

We must not accept outliers in a
calibration data set. We also can test
this with an statistical outlier test.
Thereby we have to consider that the
outlier test assumes the chosen
approach for the regression function to
be correct. First we should have a look
on the plot of the residual analysis,
because from there we can recognise
potential outliers. We calculate the
regression both with and without the
potential outlier. Then we can apply either the F-test or the t-test.

Slide 26

With the F-test we can look for significant differences in the residual
standard deviations, as shown in the
slide

Slide 27

Or we may calculate the prognostic interval of the regression function without the potential outlier, i.e. the interval within which we would expect an additional value to lie with a certain confidence.
If our suspect value lies within this interval, we cannot reject it as an statistical outlier.

Basics of Calibration

Outlier Test using the t-Test

- Calculate the prognostic interval of the regression line _without_ the potential outlier

$$PI(\hat{y}_a) = \hat{y}_a \pm t \cdot s_{y_x} \sqrt{1 + \frac{1}{N_k} + \frac{(x_a - \bar{x})^2}{\sum x_i^2 - \frac{1}{N_k}(\sum x_i)^2}}$$

$$= a_1 + b_1 \cdot x_a \pm t \cdot s_{y_x} \sqrt{1 + \frac{1}{N_k} + \frac{(x_a - \bar{x})^2}{\sum x_i^2 - \frac{1}{N_k}(\sum x_i)^2}}$$

t = tabulated value from the t distribution (P = 95%, f = N_k-2)
N_k = N-1
x_a = concentration of the potential outlier
x̄ = mean of all x_i (without x_a)

- If the potential outlier is located within the prognostic interval, it's no outlier
- If an outlier is identified statistically, the error source **has** to be found and eliminated. Than the **complete** calibration has to be repeated.

Slide 28

What we have seen up to here is the basic calibration that delivers method characteristics for the pure physical measurement in the validation and re-validation procedure. Many analytical methods however require a frequent, sometimes daily calibration. Of course there is no need to have a 10 point calibration for everyday calibration. The choice of the strategy used is the responsibility of the laboratory. It depends on the customer's requirement on accuracy and the stability of the method.

Calibration Strategies in Routine Analysis

- The basic calibration as described up to here is part of the (re-)validation of an analytical method
- For routine use calibration strategies with less effort are used
- The effort made depends on the demands of the customer and the stability of the method

Slide 29

We can choose both on the number of calibration standards and on the frequency of calibration. The one extreme for very stable methods is to use one calibration over several months. If we want we can check of course the validity of this calibration with one calibration standard.
If the method and the instrument are less stable, there might be a need for more frequent full calibrations with several calibration standards.

Calibration Strategies in Routine Analysis

- Number of calibration points in routine
  - Where the calibration of the method is very stable a one-point calibration to verify a previous multi-point calibration may be sufficient
  - In other, less stable circumstances a 3- or 5-point calibration may be needed
- Frequency of calibration
  - Also depends on the stability of calibration
  - Some analytical methods need a daily calibration whereas other calibrations may last for months. At least check of the calibration is adviceable in any case

Slide 30

In case where we have sample preparation steps with incomplete recovery
the use of an internal standard is
advisable. We have to look for a substance which is so similar to the
analyte that the chemical behaviour is
close to that of the analyte. We add a
known amount of this internal
standard to our original sample and
determine the recovery. Assuming that
we have the same recovery for our
analyte we correct our result with this recovery.
In mass spectrometry isotope marked analytes are ideal internal standards and
therefore quite often used.

Internal Standard
- Advisable for methods including a complex sample preparation procedure like extraction and clean-up
- Addition of a known amount of a substance different from the analyte and not present in the sample, but chemically behaving in the same way as the analyte
- Correction of the measurement result for the analyte with the recovery rate of the internal standard
- In mass spectrometry isotope marked analytes are often used for this purpose

Slide 31

The lower end of the working range is
limited by the detection capabilities of
the analytical method. Limit of detection and limit of quantification (sometimes also called limit of quantitation
or limit of determination) describe
these capabilities of the method.

Limits of Detection, Quantification
- Limit of Detection
- Limit of Quantification (Quantitation, Determination)

Slide 32

As the name implies the LoD
describes the lowest concentration that
can be *detected* with a certain level of
confidence. From here we can state
that the analyte is really present. This
does not mean that we can quantify the
content of the analyte with a certain
confidence. This is only possible at
concentrations above the LoQ.

Limits of detection, quantification

Limit of Detection (LoD)
Limit of Quantification (LoQ)
- Are used to quantify detection and quantification capabilities (the lower end of the working range)
- Limit of detection (LoD)
  - The lowest concentration that can be detected with a certain level of confidence
- Limit of quantification (LoQ)
  - the minimum content that can be quantified with a certain confidence

Slide 33

Most analytical procedures do not consist of a physical measurement step only, but may include chemical sample preparation steps. In addition there might be some substances present that interfere with our measurements and other constituents (the matrix) may influence the sensitivity of the method. If we simply determine the interference free or instrument detection limits, this does not describe the detection and quantification capabilities of our complete analytical method.

Limits of detection, quantification

Specification of Measurement Process

- LoD and LoQ cannot be specified in the absence of a fully defined measurement process including interferences and type of sample matrix
- "Interference free detection limits" and "Instrument detection limits", for example, do not specify the measurement capabilities of a complex measurement process including sample preparation

(IUPAC Orange Book)

Slide 34

There are many ways to calculate LoD values. Unfortunately there is no standardised way accepted by the scientific community in all fields of analytical chemistry.
We want to have a closer look on two methods, both described in the IUPAC "Orange Book".

Limits of detection, quantification

Calculation of LoD

- There is no uniform way to calculate LoD values in the scientific community
- Two possibilities are described in the following slides (both taken from IUPAC Orange Book)

Slide 35

The simple approach is just using the mean value of several determinations of blank samples plus a multiple (factor k) of the standard deviation of these measurements. With the choice of k we define the level of confidence.

Limits of detection, quantification

LoD – Simple Approach

- The value of the LOD is given by

$$LoD = \bar{x}_{bl} + k \cdot s_{bl}$$

with
$\bar{x}_{bl}$ = mean of blanks
s_{bl} = standard deviation of blanks
k = numerical factor defining the confidence level

Slide 36

We have to make a sufficiently high number of blank measurements. If the measurement of blanks does not deliver a signal we set $\bar{x}_{bl} = 0$ and use the standard deviation of measurements with a very low content. IUPAC recommends the use of k=3. Since we cannot expect that the signals of very low concentrations are normally distributed and since we only can estimate the mean and the standard deviation from a few measurements the confidence level for k=3 is expected to be around 90%.

Limits of detection, quantification

LoD – Simple Approach

- The mean of measurements of blanks and their standard deviation must be found experimentally by making a sufficiently large number of measurements
- A value of 3 for k is strongly recommended
  - At low concentrations non-Gaussian distributions are more likely
  - The mean and standard deviations are only estimates of the population characteristics
  - So the 3 s_{bl} value usually corresponds to a confidence level of about 90%

Slide 37

Now we want to have a closer look on the situation. We have to consider and avoid two different error possibilities. We want to exclude to think that the analyte is present where it is indeed not. This would be a false positive answer, a type I error.
But on the other side we also want to avoid errors, where we say the analyte is absent and it is indeed present, a false negative or type II error.

Limits of detection, quantification

LoD – More Sophisticated Approach Fundamentals

- Two kind of errors have to be considered
  - Type I – false positive
    - Probability of a type I error is called α
  - Type II – false negative
    - Probability of a type II error is called β

Slide 38

From the distribution of signals of blank measurements we can derive the *critical value*. This value is defined by the one-tailed error probability α of the statistical distribution. If the signal is above this critical value we have a low probability (α) for a false positive error.

Limits of detection, quantification

LoD – More Sophisticated Approach Critical Value

- The critical value is derived from the statistical distribution of the signal of measurements of blanks
- If the corresponding signal is exceeded we recognize with an error probability (one-tailed) of α that the analyte content in the sample is higher than in a blank
- We have a low probability for a false positive error

Slide 39

This low probability is shown in the left of the two distributions in this graph, the statistical distribution of blank measurements. At a signal corresponding to the critical value we have the low probability for a false positive error, described by the black part of the distribution.

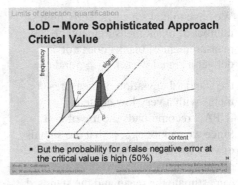

If the analyte is really present at a concentration of the critical value this means, that in 50% of the measurements we would find a signal above the critical value and in 50% a signal below the critical value.

Slide 40

But since we want to define the LoD in such a way that the probabilities both for a false positive and a false negative error are low, we have to find a value where also the probability for a false negative error β is low. IUPAC recommends to choose $\alpha=\beta=0.05$.

Slide 41

In the graph now an additional distribution of values is shown where we have only a low probability (black end of the distribution) to get a signal below the critical value. Now the probabilities for both error types are low and we can accept the mean of this population as LoD.

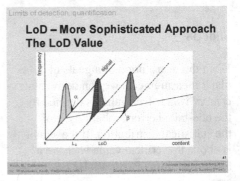

Slide 42

Let us compare the different meanings of the critical value and the LoD. If we measure a signal above the critical value we know (with a high probability) that the analyte really is present. So we can use this value as a decision limit for the presence of the analyte. The LoD gives us the minimum content where we can be sure that we really detect the analyte with a high probability. If we want to state a maximum guaranteed content ("< X") we can only use the LoD for that purpose.

> Limits of detection, quantification
>
> **Critical Value ↔ LoD**
>
> - The critical value is the decision limit for the presence of the analyte
> - The LoD describes the minimum content which can be detected with stated high probability
> - Only the LoD, not the critical value may be used as maximum guaranteed content of the analyte

Slide 43

Now we want to calculate both values from the statistical distribution of blank values or very low contents. We assume normal distribution and homogeneity of variances in the range between the blank and the LoD.
The one-tailed 95% limit for a normal distribution is at $\mu + 1.64\,\sigma$ (see also the chapter 8). We estimate μ from the mean of several blank measurements and σ from the standard s_{bl} of these measurements. If we decide to use $\alpha=\beta=0.05$, we get the formula for L_C and LoD shown in the slide.

> Limits of detection, quantification
>
> **Calculation of L_c and LoD**
>
> - If we assume a Gaussian distribution of the results and homogeneity of variances over the concentration range the limit for an one-tailed error probability α of 0.05 is at $\mu + 1.64\,\sigma$.
> - If we decide for $\alpha=\beta=0.05$ we get
> $$L_c = \bar{x}_{bl} + 1.64 \cdot s_{bl}$$
> $$LoD = \bar{x}_{bl} + 3.28 \cdot s_{bl}$$

Slide 44

There is another simple approach to define the detection limit, mainly used for chromatographical methods. The noise of the signal without a peak is quantified and a signal-to-noise ratio of 3 (sometimes 2) is used as detection limit.

> Limits of detection, quantification
>
> **Signal-to-Noise Ratio**
>
> - In many analytical applications a signal-to-noise ratio of 3 (sometimes 2) is used as definition of detection limit

Slide 45

Now we turn from detection to quanti-fication of the analyte. The slide shows the definition of the LoQ from the IUPAC "Orange Book". Note the word "adequately" in the sentence. That means that we have to define what is adequate, what is fit for the purpose.

Limits of detection, quantification

Limit of Quantification (LoQ)

- Quantification limits are performance characteristics that mark the ability of a analytical method to adequately "quantify" an analyte (IUPAC Orange book)

Slide 46

Generally a relative standard deviation of 10% is regarded as being adequate. Therefore the LoQ can be calculated as the k-fold multiple of the standard deviation at the LoQ. IUPAC recom-mends to use k=10 (corresponding to a relative standard deviation of 10%). If we assume homogeneity of variances in this very low concentration range the LoQ is $10 \cdot s_{bl}$.

Limits of detection, quantification

LoQ

- The ability to quantify is generally expressed in terms of the signal or analyte (true) value that will produce estimates having a specified relative standard deviation (RSD), commonly 10 %

$$LoQ = k \cdot s_{LoQ}$$

with

s_{LoQ} = standard deviation at the LoQ concentration

k = multiplier (reciprocal = selected quantifying RSD)

- The IUPAC default value for k is 10
- If the variances are homogeneous $s_{LoQ} = s_{bl}$

Slide 47

The *standard addition method* is a calibration *in* the sample. Known amounts of analyte are added to the samples and the signal-concentration regression line is extrapolated to a signal of zero.

Standard Addition Method

- Standard addition is calibration in the real sample by stepwise addition of known amounts of the analyte

Slide 48

The standard addition methods is a very good tool for samples where the sensitivity of the method is strongly influenced by other sample constituents. Especially where no matrix adjusted external calibration is available the benefits of the standard addition method are obvious.
If we have to analyse only a few samples the effort for the standard addition method is nearly the same as for an external calibration, but it gives us more confidence.

> **Standard addition method**
>
> ### When should the Standard Addition Method be Used?
>
> - If the composition of the sample matrix has high influence on the accuracy of analysis
> - If no matrix-adjusted calibration standards are available
> - If only few samples have to be analysed

Slide 49

There are a few requirements for the application of the standard addition method. The analytical results have to be corrected for blank. Otherwise we would add the blank value to our sample content. Since we are using linear regression we need a linear relationship between signal and con-centration. As stated above the homogeneity of variances is also a prerequisite for linear regression. We want to divide our sample into several sub-samples and spike them with known amounts of analyte. This means that we need to divide the sample homogeneously and to precisely add the analyte.

> **Standard addition method**
>
> ### Requirements
>
> - Analytical results that are corrected for blank and background
> - Linear relation between concentration x signal y
> - Homogeneity of variances
> - Possibility to homogeneously divide samples into sub-samples
> - Analyte can precisely be added to the sample

Slide 50

The procedure starts with dividing the sample into n sub-samples. We spike n-1 sub-samples with the analyte in equidistant steps and measure all n sub-samples. We use least-square regression to calculate the regression line and extrapolate to the intersection with the x-axis, i.e. we assume that our regression line is shifted to the left by the signal of our sample content. The x-axis intercept therefore delivers our sought content.

> **Standard addition method**
>
> ### Procedure
>
> - Take n sample aliquots
> - Add linear increasing amounts of the analyte to (n-1) sample aliquots in equidistant steps
> - Apply linear least square regression to the pairs of values
> $\rightarrow y = a + b \cdot x$
> - Extrapolite to the intersection with the x-axis
> - This value delivers the sought content
> $\rightarrow x_A = -x_{(y=0)} = -a/b$

Slide 51

Here we see an example. The original sample and four spiked sub-samples were analysed. The regression line intersects the x-axis at $-x_A$. So x_A is our sample content.

The advantage of the standard addition method is, that it delivers accurate results even if the recovery is not complete, provided it is constant for all sub-sample determinations.

In the graph also the confidence interval of the regression line is shown.

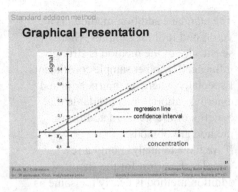

Slide 52

This confidence interval can be used to quantify the uncertainty of the result as far as it is resulting from the calibration procedure. The formula given in the slide shows the calculation of that confidence interval.

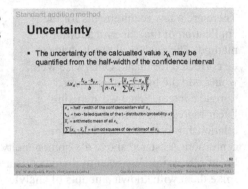

Bibliography

Funk W, Dammann V, Donnevert, G (2006) Quality Assurance in Analytical Chemistry. 2nd edition, Verlag Wiley VCH Weinheim

ISO 8466-1:1990 – Calibration and evaluation of analytical methods and estimation of performance characteristics – Part 1: Statistical evaluation of the linear calibration function

ISO 8466-2:2001 – Calibration and evaluation of analytical methods and estimation of performance characteristics – Part 2: Calibration strategy for non-linear second-order calibration functions

IUPAC (1997) Orange book – "Compendium on Analytical Nomenclature", 3rd edition, available from www.iupac.org

10 Metrology in Chemistry and Traceability of Analytical Measurement Results

Ioannis Papadakis and Bertil Magnusson

This gives an overview of metrology in analytical chemistry as well as describing the role of measurement traceability. The traceability part of the presentation is based on the CITAC position paper on measurement traceability (Traceability in Chemical Measurement, Accred. Qual. Assur. (2000) 5:388-389)

Slide 1

First we will discuss today's needs for metrology, then the fundamentals of metrology will be presented followed by some information about the international measurement system and the presentation will end with discussion about traceability of analytical measurement results.

Index

- *Metrology*
  - need for measurement quality
- International Measurement System
- Traceability of measurement results

Slide 2

Starting with the ISO definition of metrology: metrology is the science of measurement.
It is important to note that there is no distinction between theoretical or practical aspects or level of uncertainty or in whatever field of science or technology the measurements occur.

**Metrology –
Science of Measurement**

- Metrology includes all theoretical and practical aspects of measurement, whatever the **measurement uncertainty** and field of application.
  (VIM, 3rd edition)

Slide 3

But what is a measurement?
Measurement is obtaining a measurement result - a quantity value and an uncertainty.

What is a Measurement ?

Process of experimentally obtaining one or more quantity values

Quantity is a property which has a magnitude that can be expressed as a number and a unit e.g.

- Quantity: *Cadmium (mass)concentration*
- Quantity value: *12 mg/l Cd*
- Measurement result: *12 ± 2 mg/l Cd*

(VIM, 3rd edition)

Slide 4

And what is measurement quality?
Main focus is that measurement need to be comparable over space and time and here we need metrology.

What is Measurement Quality ?

- Results should be fit for purpose – regarding several parameters e.g uncertainty, price and comparability

Comparability - measurements need to be comparable over:

- Time   1900   2000   2100   year
- Between different laboratories
- Between different countries

measured once – accepted everywhere

Slide 5

It should also be clear that unreliable measurements lead to duplication of work, use of extra resources, lack of trust between trading partners, create negative economic impact and often lead to accidents or disasters.

Lack of Measurement Quality can Lead to:

- Duplication of measurements
- Use of extra resources
- Lack of trust
- Negative economic impact
- Disasters/accidents

Slide 6

Let's see now some examples, which show the importance of measurements for today's society.
The first example (picture) is relevant to physical metrology.

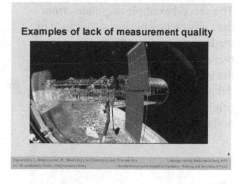

Examples of lack of measurement quality

Slide 7

The picture in the previous slide is a scientific space ship. Another scientific spaceship, the Mars Climate Orbiter, was completely destroyed on September 1999. The reason for this disaster was that two different NASA teams were using different systems of units. Somewhere in the conversion a mistake was made that caused the loss of the spacecraft. This probably would have been avoided if the same system of units had been used.

Mars Climate Orbiter

...confusion about units leads to crash...

- On 23 September 1999 the Mars Climate Orbiter, one of the missions in a long-term program of Mars exploration, burned out completely.
- The accident was not due to a technical problem, but the result of the different measurement units used by the NASA teams.
- One team used the metric system for spacecraft operation, while the other used the English units. This information was critical to the maneuvers required to position the spacecraft in the proper Mars orbit and led to the loss of the orbiter.

The fate of the Mars Climate Orbiter clearly shows the need for standardization of units

Slide 8

In order to make it more explicit for chemical measurement, three examples are chosen from different chemical measurement fields, which show the importance of chemical measurement. The first originates from petroleum industry. A single measurement mistake cost 10 million US dollars to the company.

Difference $ 10 Million

- A subsidiary of an oil company in the Far East analysed a batch of petrol. Their local lab established that the gum content (components in the gasoline that polymerize during combustion) was much too high.
- On the basis of this analysis the company sold the batch to a trader for a much lower price.
- The trader asked a second lab to perform an analysis in order to find out what he could do with the off-spec petrol.
- He was very pleasantly surprised to find that the gasoline was actually on-spec and he was able to make a healthy profit selling the batch for the normal price.

The oil company only found out much later that the problem was not the petrol, but an error at their own lab. By then this error had already cost them $ 10 million.

(Buskes and van Gerven)

Slide 9

The second example originates from the construction business. Because of inadequate quality control measurements on the construction materials the whole project was considerably delayed and the total cost, which was originally estimated as 900 million US dollars, exceeded 1 billion US dollars.

Alaska Pipeline

- The 800-mile trans-Alaska pipeline pumps oil from the northern coast to the southern border of Alaska.
- Construction started in 1973 and was completed 4 years later.
- The pipeline was originally budgeted $ 900 million, but the cost escalated to exceed $ 1 billion.
- A steel manufacturer was awarded the multimillion dollar contract to supply steel for the pipeline with S content of less than 0.005%.
- When several of the joint welds in the pipeline began to fail, it became clear that the S content was much higher than specified.

The poor quality of the steel, in part due to inadequate or lack of measurements, set the project back several millions of dollars, once again emphasizing the need for accurate measurements.
(Buskes and van Gerven)

Slide 10

The third, and final, example originates from the health sector.
A detailed report in the USA showed that by reducing the uncertainty of cholesterol measurement from 24% to 5%, the USA economy saved, every year, 100 million US dollars.
This arose from avoiding unnecessary treatment and improving the original diagnosis.

Cholesterol Measurements

- A high measurements uncertainty for cholesterol can lead to an unnecessary costly treatment or a higher health risk.
- Reducing the measurement uncertainty from 23.7% in 1949 to 5% in 1995, saves to the Unites States alone $ 100 million every year in health care costs.

Standard reference materials played an important role in lowering the measurement uncertainty.
(W. May)

Note: Today other compounds (lipoproteins) are used for risk markers of myocardial diseases

Slide 11

After these examples it should be clear now that measurements are very important with, amongst others, economic, political, social, environmental and scientific impacts.

Importance of Measurements

- Economic
- Political
- Social
- Environmental
- Scientific

Slide 12

Today metrology is used by practically everybody.
We see in this slide a list of the main users of metrology.

Measurements are used in

- Industry (e.g. manufacturing)
- Commerce (e.g. disputes)
- Governments (e.g law implementation)
- Health and safety
- Environment protection
- Science / research
- Military services (e.g. navigation)
- Communications

Slide 13

Let's see how metrology started!
Before the 19th century, the measurement unit was something very unequivocal.
In most cases the units depended very much on the 'locality' and very often were related to something that appeared to be constant like the king's foot!

... Lack of Standard ...

King's foot

Slide 14

An interesting example was the length unit 'ell'. This unit was used in many countries in Europe, but the actual length was different in different places.

... Lack of Standard ...
Variations of One Unit of Length (Ell)

- The "ell", a unit originating from the custom of measuring cloth using one's forearms, existed in many countries.
- In order to make trade possible at all in these days, conversion tables were used.
  (Buskes and van Gerven)

Slide 15

This table shows the length (in meters) of the unit 'ell' in different countries and cities in the middle ages.

... Lack of Standard ...

country	ell(m)	city	ell(m)
England	1.14	Vienna(A)	0.78
Scotland	0.94	Bruges (B)	0.70
Germany	0.6	Amsterdam (NL)	0.69
Russia	0.5		

(Buskes and van Gerven)

Slide 16

This lack of standards was resolved in the second half of the 19th century with the signature of the Meter Convention. This convention is a diplomatic treaty, which was signed in Paris (France) on 20th May 1875. This treaty establishes the International System of Units (SI) for the signatory countries. Currently 52 countries (all the major industrialized countries) have signed the treaty, and 36 countries are associate members.

Meter Convention

- Diplomatic treaty
- 20th May 1875, in Paris
- SI system
- 52 signatory countries
- 36 associate members

Slide 17

The main aims of the Meter Convention are to achieve international uniformity of measurements of all kinds, establish a common system of units, achieve equivalent measurement standards in the members states, harmonize laws and regulations relative to measurements in the member states and achieve mutual recognition of measurements.

Meter Convention Aims

- International uniformity in measurement
- Common system of units
- Equivalent measurement standards
- Harmonised laws and regulations
- Mutual recognition of measurements

Slide 18

As mentioned before, the SI is the major instrument of the meter convention.

Slide 19

The SI consists of seven base quantities from which all the other quantities (secondary or derived quantities) can be derived. The table in the slide presents the name of the quantities, their unit and the symbol of the unit.

SI Base Quantities

quantity	unit	symbol
▪ Length	metre	m
▪ Mass	kilogram	kg
▪ Time	second	s
▪ Electric current	ampere	A
▪ Thermodynamic temperature	kelvin	K
▪ Amount of substance	mole	mol
▪ Luminous intensity	candela	cd

Slide 20

Some examples of derived quantities are given in this slide. In particular the last one is widely used in analytical chemistry.

SI Derived Quantities Examples

quantity	unit	symbol
▪ Speed, velocity	metre per second	m/s
▪ Density	kilogram per cubic metre	kg/m^3
▪ Concentration (of amount of substance)	mole per cubic metre	mol/m^3

Slide 21

What about chemistry and the SI? Chemistry is a quite new discipline for the SI.
The quantity relevant to chemistry (i.e. the amount of substance) was adopted only in 1971, almost 100 years after the original signature of the meter convention. The unit of the amount of substance in the SI system is the mole (mol).

Chemistry in SI
It is quite new!

- Amount of substance (AoS)
- Agreed on 1971
- Mole (mol)

Slide 22

What about traceability? Traceability is the relationship of a result of a measurement to a value of a standard through an unbroken chain of comparisons (traceability chain). In the case of the length measurements this chain can be realized in the way presented in the slide. When we want to measure the length of a shark, the result of our measurement is dependent on the tape we are using (comparison).

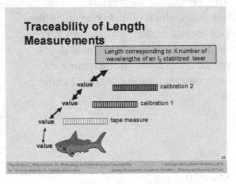

The scale of the tape depends on the calibration (comparison) of the production instrument and this depends on a number of other calibrations. The end of this 'chain' is the highest calibration, which is a comparison against the SI standard, in other words the realization of the meter.

Slide 23

There is often a misunderstanding related to traceability. It is often believed that the aim is traceability itself, but this is not true. The actual aim is to obtain comparable measurements in time and space. Traceability is just the best possible way to achieve this.

analytical measurements
need to be comparable
in time and space

traceability *is the best*
way to achieve this

Slide 24

Now let's see how a traceability chain can be realized for a chemical measurement. Generalizing and in an analogous way to length measurements it can be like the chain in the slide. The amount content of a compound X in a solution is compared with the amount content in a working standard. This in turn was compared with the amount content in a reference standard and after some further comparisons, in an ideal situation we end up with the SI unit, the mole.

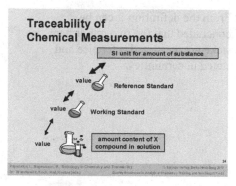

Slide 25

This model of course makes a lot of assumptions and there are quite a few problems. It is not possible to have appropriate reference standards for all possible chemical measurements, very often there are no common links (i.e. reference to a common basis, which is ideally the SI), laboratories do not always use the standards in an appropriate way and very often the uncertainty concept is not used at all!

Problems...

- Absence of reference standards
- Absence of links to common basis
- Appropriate use of standards by laboratories
- Appropriate use of uncertainty

Slide 26

Before continuing, let's see the official definition of traceability as given in the ISO international vocabulary of basic and general terms in metrology. Traceability is defined as the property of the result of a measurement or the value of a standard whereby it can be related to stated references, usually national or international standards, through an unbroken chain of comparisons all having stated uncertainties.

Traceability - Definition

Metrological traceability
Property of a **measurement result** whereby the result can be related to a reference through a documented unbroken chain of **calibrations**, each contributing to the **measurement uncertainty**

(VIM, 3rd edition)

Slide 27

From the definition it can be concluded that the two major items at stake are the stated reference and stated uncertainties.

Slide 28

Various stated references are involved in a chemical measurement. In VIM the three different references possible are given with examples. For determining amount of substance several references can be involved e.g. we could weigh a material with known purity and from the known molar weight we could calculate the amount content. All the references used in a measurement should be stated and taken appropriately into account.

Stated References – 3 different

- In VIM 3 examples of different stated references are given
  - A measurement unit (VIM 1.9), e.g. mol/l, °C
  - A measurement standard (VIM 5.1), e.g. the certified reference material SRM 2193, a $CaCO_3$ pH standard.
  - A measurement procedure (VIM 2.6), e.g. ISO 1736:2008 Dried milk ... - Determination of fat content
- Determination of amount of substance requires in most cases measurements of different properties
  - Sample mass   mass reference – measurement unit
  - Analyte identity   pure material – measurement standard
  - Molar or Atomic weight published data or measured

Slide 29

The evidence required by the laboratory to demonstrate traceability for the mercury result is here given.
Point 1 would need special attention to assure the quality and traceability for the calibration standard.
Traceability for points 2, 3 and 6 is easily achieved with adequate uncertainty using commercial equipment.
Points 4 and 5 need additional attention by the lab.

Several References for one measurand

For measurements with more than one input quantity in the measurement model, each of the input quantity values should itself be metrologically traceable...
NOTE 4 in VIM on Traceability

Example: Mercury in tuna fish (with a AAS after microwave digestion)
Measurement result: 4.03 ± 0.11 mg/kg, reported as total Hg on dry weight basis (105 °C, 12 h)
Traceability has to be demonstrated for:
- Mass concentration of the Hg solution **1.00 g/l Hg** - a CRM certificate
- mass of sample **0.5 g** - calibration certificate of the balance
- volume of volumetric flask **100 ml** - calibration certificate
- drying temperature **105°C** - calibration of oven
- drying time **12 h** - ordinary clock or stopwatch
- Microwave digestion conditions **0.5 h at 180 °C** - check according to specifications

(from Eurachem Traceability leaflet - www.eurachem.org)

Slide 30

The case of the stated uncertainties is quite different. Uncertainty is the interval around the measurement results, which should contain the so-called 'true value'. The uncertainty statement should always be accompanied by an uncertainty budget, which should include *all* uncertainties carried by all of the references used as well as the uncertainties introduced by the measurement process.

Stated Uncertainty

- An interval around the measurement result
- The uncertainty budget including:
  - Uncertainties carried by the references
  - Uncertainties introduced by the measurement process

Slide 31

In general the uncertainty contributions carried by the references are quite small compared to those, which originate from the measurement process.

Stated Uncertainty

Usually the contribution of the uncertainties carried by the references to the total uncertainty is small relative the contributions that originate from the measurement process

Slide 32

As far as reference materials are concerned, the property values carried by reference materials, should be traceable to other references and in an analogous way the same features which are valid for the analytical laboratories, are also valid for the reference materials producers.

Reference Materials

- Values carried by reference materials should be traceable to other references
- The same features which are valid for the analytical laboratories are also valid for the reference materials producers

Slide 33

Detailed information relating to
metrology in chemistry and traceability
can be obtained from these web-sites.
The addresses include international
and regional organizations as well as
some national metrology institutes.

More Information

- www.bipm.org
- www.euramet.org
- www.citac.ws
- www.eurachem.org
- www.eurolab.org
- www.irmm.jrc.be
- www.nist.gov
- www.labnetwork.org

Slide 34

Summarizing this chapter, you must
remember that metrology in chemistry
is still a young discipline thus there is
still a lot to be learned, traceability is
not an aim by itself, but it is the way to
achieve reliable and comparable
results. Traceability can only be
claimed when the uncertainty statement includes all the uncertainty
contributions from the references and
the measurement process.

Things to Remember

- Metrology in chemistry is still "young"
- There is a lot to learn
- *Traceability is not an aim by itself but it helps achieving reliable results*
- *Traceability can only be claimed if uncertainty statement includes all the uncertainties from references and the measurement procedure*

Bibliography

Buskes H, v.Gerven B (1999) Footprint of the Meter. NMi Communicabus

EURACHEM / CITAC (2008) Metrological Traceability of Analytical Results.
Leaflet available from www.eurachem.org

ISO/IEC Guide 99:2007 International Vocabulary of Metrology – Basic and
General Concepts and Associated Terms (VIM), 3rd edition, also available as
JCGM 200:2008 from www.bipm.org

Papadakis I, Wegscheider W (2000) CITAC Position Paper: Traceability in
Chemical Measurement, Accred Qual. Assur 5:388-389

May W (1999) NIST SRMs. What they are, what they are NOT and their impact on the U.S. economy, international trade and quality of life. CITAC '99 Japan Symposium on Practical Realization of Metrology in Chemistry for the 21th Century, Nov 9-11, Tsukuba, Japan

11   Validation of Analytical Methods – to be Fit for the Purpose

Bernd Wenclawiak and Evsevios Hadjicostas

Fitness for purpose is the ultimate goal of the person doing the job in the laboratory, especially for the choice of method and instrumentation used to carry out an analysis. The validation of the analytical method is the important part to guarantee the fitness.

This chapter aims to assist laboratories in implementing the principle that analytical measurements should be made using methods and equipment that have been tested to ensure they are fit for purpose.

Slide 1

The following transparencies are a series of statements. Each emphasizes one part of the important role of analytical measurements. They go without saying to an analytical chemist. Satisfaction of requirements is scale of quality.

Analytical measurements should be made to satisfy an agreed requirement

Slide 2

This is the second one. If the method and the equipment are not fit for purpose there is no chance to deliver good quality.

Analytical measurements should be made using methods and equipment which have been tested to ensure they are fit for purpose

B.W. Wenclawiak et al. (eds.), *Quality Assurance in Analytical Chemistry: Training and Teaching*, DOI 10.1007/978-3-642-13609-2_11, © Springer-Verlag Berlin Heidelberg 2010

Slide 3

This is the third one. This statement is important. It refers to people. The staff doing the work have to be qualified and competent. This can be achieved by recruiting suitably qualified staff or by in service training. Competence arises from experience and training. A supervisor has to make sure that the staff are well trained and competent.

Staff making analytical measurements should be both:
qualified and competent
to undertake the task and demonstrate that they can perform the analysis properly

Slide 4

This is the fourth statement. Laboratories must participate in regular independent assessments to demonstrate their technical performance.

There should be a regular independent assessment of the technical performance of a laboratory

Slide 5

This is the fifth statement. This implies, in many cases, that there must be an agreement on the method selected for the given analysis or, even better, the measurement is traceable to an agreed reference (see chapter 10).

Analytical measurements made in one location should be consistent with those made elsewhere

Slide 6

This is the sixth statement. Organizations making analytical measurements should have well defined quality control and quality assurance procedures. These procedures are explained in detail throughout this book.

Organizations making analytical measurements should have well defined quality control and quality assurance procedures

Slide 7

Validation is confirmation, through the provision of objective evidence that the requirements for a specific intended use or application have been fulfilled (ISO 9000:2005). Method Validation is therefore the process of confirmation that a method is fit-for-purpose, i.e. suitable for solving a particular analytical problem.

What is Method Validation?

Method validation is an important requirement in the practice of chemical analysis

Method validation is the process of proving that an analytical method is acceptable for its intended purpose

Slide 8

This process consists of the establishment of the performance characteristics (e.g. precision, trueness, uncertainty) and the limitations of the method (e.g. limit of detection, limit of quantification, working range). It is also important to identify the influences that may change these characteristics (e.g. interfering substances, changes in the sample matrix) and to what extend they might change them. Which analytes can be determined, in which matrices and in the presence of which interfering substances? Within these conditions: what levels of precision and accuracy can be achieved? Is the measurement uncertainty of the result acceptable for the intended use?

What is Method Validation?

- The process of establishing the performance characteristics and limitations of a method and the identification of the influences which may change these characteristics and to what extent.
- Which substances can it determine in which matrices in the presence of which interferences?
- Within these conditions what levels of precision and accuracy can be achieved?

This process includes questions like:
- Is there a linear relationship between the signal and the concentration of the analyte?
- Are higher order terms required?
- Over what concentration range is this relationship valid? This means for example that the signal obtained from an instrument should only be the signal arising from the analyte, i.e. that there is no interference from other analytes present in the same sample/matrix.
- What precision and accuracy can be achieved?

Just compare a simple photometric experiment with atomic absorption measurement and check for the effects from possible interferences.

Slide 9

This is what analytical chemistry is all about, the selection from the methods that are available, the one which is most suitable for the task i.e. a method that is fit for purpose. For example you would probably never choose NMR for the determination of cations in water (ICP-OES or AAS are certainly more suitable).

The Method Validation Process Implies:

- The process of verifying that a method is fit for purpose, i.e. for use for solving a particular analytical problem
- *This means the method must be suitable*
- *Method validation is not solely the process of evaluating the performance parameters*

Slide 10

The equipment and instrumentation used for the determination of the performance parameters must be
- within specification,
- adequately calibrated and
- working correctly.

Likewise the operator carrying out the studies must be
- competent in the field of work under study and

The Method Validation Process Implies:

- Method performance parameters are determined using equipment that is:
  - Within specification
  - Working correctly
  - Adequately calibrated
- Operator is fit for purpose
- Method validation and method development

- have sufficient knowledge related to such work to be able to make appropriate decisions. Such decisions are made after appropriate comparisons and statistical interpretation of the results.

Many of the method performance parameters are usually evaluated as part of method development. It is for this reason that method validation is very closely related to method development.

Slide 11

The reason why method validation is necessary comes from the fact that measurements are directly connected to activities of everyday life. The cost of carrying out these measurements is high and additional costs –quite often even higher - arise from decision made on the basis of the results. Therefore it is necessary to have the confidence of the interested parties. It is for this reason that the performance para-meters of the method are validated and the uncertainty of the results is evaluated at a predefined confidence level. It is to be noted that most of the information required to evaluate uncertainty can be obtained during the validation process or routine quality control (see chapter 12).

Why is Method Validation Necessary?

- To increase the value of test results
- To justify customer's trust
- To trace criminals
- To prove what we claim is true
- Examples
  - To value goods for trade purposes
  - To support health care
  - To check the quality of drinking water

Slide 12

When performing laboratory work the analytical chemist should know that the analytical results produced are not numerical figures with no further meaning. Most of the time the labora-tory results have a considerable and multi-fold impact on everyday life. This impact affects the economic, social, juristic, welfare, environmental and many other interests of the labora-tory's customers. Therefore, the professional duty of the analytical chemist is to respect the interests of all the interested parties and to carry out his work with integrity in order to contribute effectively to the quality of life.

The Professional Duty of the Analytical Chemist

- To increase reliability of laboratory results
- To increase trust of laboratory customers
- To prove the truth

Slide 13

A method should be validated when it is necessary to verify that the performance parameters are adequate for use for a particular analytical problem. Method validation is required in circumstances such as those indicated in this slide.

The extent to which method validation has to be decided by the laboratory, taking into consideration the time and cost to perform the validation on one hand and on the other hand the consequences that arise from a insufficient validation. The changes made in reapplying a method in different laboratories, with different instrumentation or operators, as well as changes made to meet customer requirements and relevant regulations, always have to be considered.

When should Methods be Validated?

- New method development
- Revision of established methods or instruments
- When established methods are used in different laboratories/different analysts etc.
- QC indicates method changes
- Comparison of methods

Slide 14

This is the main part of this chapter. Who? What? Why? How? Common questions in a new field.

How should Methods be Validated?

Slide 15

When a method is being developed, especially if it will be widely used, the validation is usually carried out by a group of laboratories. However, it is not always feasible for a laboratory to participate in or organize a collaborative study, since no other laboratory may be interested in taking part in a specific study. The competition between laboratories in the same or similar field of interest is also a constraint,

Who Carries out Method Validation?

- Validation in a single laboratory
  - Comparisons with CRMs
  - Comparisons with other methods that are validated
- Validation in a group of laboratories
  - Collaborative studies
  - Inter-laboratory comparisons

that prevents laboratories from participating in such studies. In each case every laboratory has to investigate its performance factors such as linearity, interferences, or robustness.

Where it is inconvenient or impossible for a laboratory to enter into a collaborative study it is possible to carry out method validation following internal procedures. In such cases, which are the most common, there is a need to provide objective evidence that the validation was carried out, using appropriate and approved methodology. The use of Certified Reference Materials or the comparison against a previously validated method is highly recommended in order for a validation to be recognized by regulatory or accreditation bodies.

Slide 16

The customer gives the analytical laboratory information about the characteristics of the sample submitted for analysis. Such information should be, for example, the analyte to be measured, the expected concentration levels, which substances that might interfere with the determination of the analyte are present, how the sample was taken etc.

How should Methods be Validated?

The Analytical Requirement (I)

- What are the analytes of interest?
- What are the expected concentration levels?
- Are there any interferences?
- How was the sampling done?

Slide 17

Faced with an analytical problem, the laboratory, if necessary in collaboration with the customer, evaluates existing methods for their suitability and makes a decision on which method probably is the most suitable, what degree of validation is required and how it will be used to solve the analytical problem in question. The decision, based on the results of the validation process, that the method is fit-for-purpose is usually left to the analyst's discretion, who also decides what performance is required, what instruments and what facilities are to be used etc.

How should Methods be Validated?

The Analytical Requirement (II)

- Which method is the most suitable?
- What degree of validation is required?
- How will the method be used?

Slide 18

Method development can start with minor modifications on an existing method or may require the development of a completely new one. If necessary, the selected method for solving an analytical problem may need further development or validation of more performance characteristics. Method development can take a number of forms. The analytical chemist may start out with a few sketchy ideas and apply expertise and experience to devise a suitable method or he/she may adapt existing methods making minor changes so that the method is suitable for a new application.

Slide 19

As a general rule, it is necessary for a laboratory to be able to have an adequate level of confidence in the results produced; otherwise the work is not worth doing. The validation of a method usually is costly and time consuming and in some cases it may be more appropriate to subcontract the work to another laboratory instead of installing a new method. If possible the customer should be involved in the decision on the degree of validation required. The knowledge of the laboratory about the performance of the method to be used as well as any data showing compatibility of the method in question with similar methods helps to decide the degree of validation required.

Slide 20

Being familiar with a method ensures acceptable performance. New operators, this might include newcomers to a job or students or other trainees, need training before performing the job on their own. First the trainee studies the instructions and handouts provided. He has to familiarise himself with the theory behind the measurement. The trainer/supervisor demonstrates the method while the trainee is watching.

Rules Recommended to Ensure Acceptable Performance

- Analyst should be familiar with method before using it for the first time
  - Work firstly under supervision
  - Get training
  - Think ahead of process, solutions etc. required
- Make an assessment how many samples can be handled at a time
- Make sure everything needed is available before work starts

After this the trainee uses the method under supervision with material analysed before, e.g. in-house reference material. Can he achieve the performance stated in the method? He has to learn to think ahead, making sure that all material/solutions are ready when required.

The assessment should give information on how many samples can be handled at a time.

Standards, reagents, and equipment (and sometimes bench space) should be available before the work starts.

Slide 21

The effort required to get the necessary information can be minimized with careful planning. This is because a particular set of experiments often yields information on several parameters. Some of the parameters may have been determined during the method development stage. The laboratory makes the appropriate decision as to the degree of validation required taking into account the customer's

Deciding what Degree of Validation is Required

Category of the method	Action necessary
Interlaboratory tested	Precision, trueness
Interlaboratory tested but it applies with different material, different instrument	Precision, trueness, limit of detection, selectivity
Established but not tested	Many
From bibliography, with reference to performance characteristics	Many
From bibliography, without reference to performance characteristics	Many
In-house method	Full validation

requirements, existing experience in the use of the method and the need for compatibility with other similar methods already in use within the laboratory or being used by other laboratories.

Slide 22

The laboratory should agree with the customer which requirements with respect to analysis and performance the method must have. Is an existing method suitable or must a new method be developed? Evaluation and development of a new method are a continuing process which is best explained in the depiction given here. This figure has been taken from the "The Fitness for Purpose of Analytical Methods: A

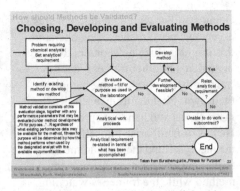

Laboratory Guide to Method Validation and Related Topics" (1998) published by EURACHEM.

Slide 23

If the signal can be attributed only to the analyte *identity* is confirmed. *Selectivity* is important, if interferences are present. For example a reference compound and an analyte compound generate a peak in a chromatogram at the same retention time. Either columns of different polarity or a mass spectrometer as detector can help to identify a single compound peak.

Limit of Detection (LoD) is the lowest concentration that can be detected with specified confidence for a specific substance.

Limit of Quantification (LoQ) is the minimum content that can be quantified with a certain confidence. Values below LoQ are reported as less than (see also chapter 9).

Accuracy is the closeness of a result to a true value and combines precision and trueness.

Trueness is the closeness of agreement between the average value obtained from a large set of test results and an accepted reference value.

The measurement *bias* is an estimate of a systematic measurement error.

Precision is the closeness of agreement between indications or measured quantity values obtained by replicate measurements on the same or similar objects under specified conditions.

Repeatability means conditions of measurement that includes the same measurement procedure, same operators, same measuring system, same operating conditions and same location, and replicate measurements on the same or similar objects over a short period of time.

Reproducibility means conditions of measurement that includes different locations, operators, measuring systems, and replicate measurements on the same or similar objects.

Sensitivity of a measuring system is the quotient of the change in an indication of a measuring system and the corresponding change in a value of a quantity being measured.

Robustness of an analytical procedure is a measure of its capacity to remain unaffected by small, but deliberate variations.

Recovery is the fraction from an analyte added to a test sample (fortified or spiked sample) prior to analysis that finally was found with the analysis. The percentage recovery (%R) is calculated as %R = $[(C_F-C_U)/C_A]$ x 100, where C_F is the concentration of analyte measured in the fortified sample, C_U is the concentration of analyte measured in the unfortified sample and C_A is the concentration of analyte added [AOAC-PVMC].

Slide 24

During the analysis process there is the stage of measuring the signal attributed to the analyte. The analytical chemist must prove whether this signal corresponds to the analyte or whether part of it is to be attributed to interfering impurities. This is the confirmation of *identity*. *Selectivity* is a measure that assesses the reliability of measurements in the presence of interferences. The Selectivity of a method refers to the extent to which it can determine particular analyte(s) in a complex mixture without interference from the other components in the mixture. The term specificity was used in the past to describe complete selectivity. Due to the fact that perfect selectivity can never be reached, IUPAC recommends, not to use this term.

How should Methods be Validated?

Identity and Selectivity

- *Identity*: Signal to be attributed to the analyte e.g. GC (use two differently coated columns), GC-MS and GC-IR (use two different detectors)
- *Selectivity*: The ability of the method to determine accurately the analyte of interest in the presence of other components in a sample matrix under the stated conditions of the test.

Slide 25

Confirmation increases confidence in
the technique under examination and
is especially useful when the confirmation techniques operate on significantly different principles. It is essential
to distinguish between the meanings of
repeatability and confirmation.
Whereas repeatability requires the
measurement to be performed several
times by one technique, confirmation
requires the measurement to be
performed by more than one technique.

Identity and Selectivity

- Confirmation versus repeatability
- Confirmation: Measure by more than one technique
- Repeatability: Measure several times by one technique

Slide 26

The selectivity of the method is
usually investigated by studying its
ability to measure the analyte of interest
in test portions to which specific interferences have been deliberately
introduced (those thought likely to be
present in samples). Where it is
unclear whether or not interferences
are already present, the selectivity of
the method can be investigated by
studying its ability to measure the
analyte by comparison with other independent methods or techniques.

Identity and Selectivity

- How to establish selectivity:

  Compare the response of the analyte in a test mixture with the response of a solution containing only the analyte.

Slide 27

The procedure to establish selectivity
of the method is the following:
-    Analyse samples and reference
     materials by the candidate and
     other independent methods.
-    Assess the ability of the methods
     to confirm analyte identity and
     their ability to measure the analyte
     in isolation from the interferences.

Identity and Selectivity

- The procedure to establish selectivity:
  - Analyze samples and reference materials
  - Assess the ability of the methods to confirm identity and measure the analyte
  - Choose the more appropriate method.
  - Analyse samples
  - Examine the effect of interferences

- Choose the most appropriate method.
- Analyse samples containing various suspected interferences in the presence of the analyte of interest.
- Examine the effect of interferences and decide whether further method development is required.

Slide 28

Where measurements are made at low concentration levels, e.g. in trace analysis, it is important to know what is the lowest concentration of the analyte that can be confidently detected by the method. For validation purposes it is normally sufficient to provide an indication of the level at which detection becomes problematic. *Limit of detection* is the lowest concentration of analyte in a sample that can be detected, but not necessarily quantified under the stated conditions of the test.

How should Methods be Validated?

Limit of Detection (LoD)

- LoD: The lowest concentration of analyte in a sample that can be detected with a certain level of confidence
  - LoD = B + 3 s_0 or 0 + 3 s_0 (for fortified samples) (typically, three times the noise level)
    B = Blank
    s_0 = standard deviation of 10 measurements

Slide 29

The Limit of Detection is expressed as the concentration or the quantity derived from the smallest signal that can be detected with reasonable certainty for a given analytical procedure. One possibility to express the LoD is as the mean blank value plus three times the standard deviation of 10 measurements of the blank sample (see also chapter 9). In the case where a blank sample does not produce a measurable signal, fortified samples on a very low level have to be used.

How should Methods be Validated?

Expression of the LoD

- Analyse
  - 10 independent sample blanks and get the mean sample blank value (B) or
  - 10 independent sample blanks fortified at lowest acceptable concentration.
- Express LoD as the analyte concentration corresponding to
  - B + 3 s or
  - 0 + 3 s
  - (s being the sample standard deviation).

Slide 30

Limit of Quantification (LoQ) is the lowest concentration of analyte in a sample that can be determined with acceptable accuracy, i.e. with acceptable precision and acceptable trueness, under the stated conditions of the test. It is an indicative value and should not normally be used for decision-making purposes. It should be established using an appropriate measurement standard or sample and should not be determined by extrapolation. The LoQ is calculated as the analyte concentration corresponding to the sample blank value plus 10 standard deviations of the blank measurement. If measurements are made under repeatability conditions, a measure of the repeatability precision at this concentration is also obtained.

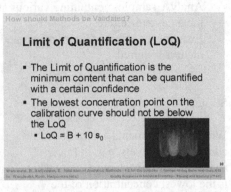

Slide 31

The calculation of the Limit of Detection and/or Limit of Quantification can also be made using a graphical method. In this case the analyst prepares at least three test samples at different concentration levels and makes at least six measurements at each concentration and calculates the mean concentration c and standard deviation s. The three mean concentration values are the independent variables and are plotted on the X-axis against the corresponding s values which are the dependant variables plotted on the Y–axis. s_0 is the intercept at the Y-axis and LoD = 3 S_0 whilst LoQ is 10 S_0.

For more background on the determination of LoD and LoQ see chapter 9.

Slide 32

Linearity is the ability of the method to obtain measuring signals proportional to the concentration of the analyte. Linear range is by inference the range of analyte concentrations over which the method gives signals proportional to the concentration of the analyte. For any quantitative method it is necessary to determine the range of analyte concentrations over which the method may be applied. At the lower end of

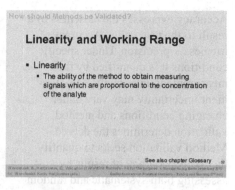

the concentration range the limiting factors are the values of the limits of detection and quantification. At the upper end of the concentration range various effects, depending on the response of the instrument system, will impose limitations.

Slide 33

The *working range* is limited by the lowest and the highest calibration point. The lowest one should not be below the LoQ. Within the working range there may exist a linear response range within which the signal response will have a linear relationship to the analyte concentration. Linearity for quantitative methods is determined by the measurement of samples with analyte concentrations spanning the claimed range of the method. The results are used to calculate a regression line against analyte calculation using the least squares method. Different criteria are used to establish good correlation of the concentration with signal. The coefficient r is a measure of correlation not a measure of linearity! Linearity can be checked by determining the signal to concentration ratio. A range below 5% of the averaged signal to concentration ratio is considered linear. Plotting the calibration range can reveal the results better than any mere calculation.

Evaluation of the working and linear ranges will also be useful for planning the degree of calibration required when using the method on a day-to-day basis. Within the linear range, one calibration point is sufficient to establish the slope of the calibration line whereas a multi-point calibration is required when working elsewhere in the working range. In general linearity checks require points at at least 10 different concentrations/property values.

Slide 34

Accuracy expresses the closeness of a result to the true value. Accuracy = trueness + precision. Under specific conditions it is quantified by the measurement uncertainty. Measurement uncertainty may vary under changing conditions and method validation determines the degree. Method validation seeks to quantify the likely accuracy of results by assessing both systematic and random effects on results. The property related to systematic errors is the trueness, i.e. the closeness of agreement between the average value obtained from a large set of test results and an accepted reference value. The property related to random errors is precision, i.e. the closeness of agreement between independent test results obtained under stipulated conditions. Accuracy is therefore, normally studied as trueness and precision.

Accuracy/Trueness/Precision

- *Accuracy:* Closeness of agreement between a measured quantity value and a true quantity value of a measurand [VIM]
- *Trueness:* The closeness of agreement between the average of an infinite number of replicate measure quantity values and a reference quantity value [VIM]
- *Precision:* closeness of agreement between indications or measured quantity values obtained by replicate measurements on the same or similar objects under specified conditions [VIM]

Slide 35

A result could have "good precision" but a "bad trueness". That is, all measurements could be repeatable, but the mean value could be far away from the true value. This is not a "good accuracy". Similarly, a result could have "good trueness" whilst precision could be bad. In this case the mean value may be close to the true value even though precision is low. This is again not a "good accuracy". See also chapter 8 for further information.

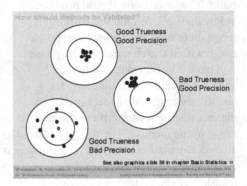

Good Trueness Good Precision

Bad Trueness Good Precision

Good Trueness Bad Precision

See also graphics slide 36 in chapter Basic Statistics

Slide 36

Trueness is assessed against a reference value (i.e. a conventional true value). Trueness is a property related to systematic errors.
Two basic techniques are available: checking against reference values
-    from a characterized material or
-    from a characterized method.
Certified Reference Materials (CRM) should provide traceable values (to international standards), which are the "reference values". Trueness can also be studied via interlaboratory studies as well as by using recovery study on spiked samples and calculating the recovery. The possibilities stated in the slide are hierarchically sorted, i.e. the top methods are usually regarded as the most reliable ones.

Determination of Trueness

- Using Certified Reference Materials
- Using RM or In-house materials
- Using Reference methods
  - Single sample
  - Many samples
- Via Interlaboratory study
- Via recovery study

Slide 37

To check trueness using a reference material, a series of replicate tests is carried out and the mean value and standard deviation are compared with the characterized value for the reference material. Certified matrix reference materials may be purchased and used directly for validation. If this is not possible reference materials for validation may be prepared by spiking typical materials with pure Certified Reference Materials or other materials of suitable purity and stability. In-house reference materials may be prepared from typical, well-characterized materials checked in-house for stability and retained for in-house Quality Control.
The trueness is determined by comparing the arithmetic mean, $\bar{x}$, of the value that the laboratory assigns to the Certified Reference Material with the reference value x_{ref} of the CRM, which is (by definition) accepted as sufficiently true. If it is found to be within the confidence interval of the laboratory's results, then this is judged to be a satisfactory outcome.

Determination of Trueness

- Using Certified Reference Materials
  - A 1-sample-t-test can be applied
  - Perform some analyses (e.g. at least five*), then calculate the confidence interval of the arithmetic mean and check if the reference value is within this interval

$$\bar{x} - \frac{t}{\sqrt{n}} \cdot s \; \langle \; x_{ref} \; \langle \; \bar{x} + \frac{t}{\sqrt{n}} \cdot s$$

*NORDTEST Handbook

Slide 38

For the evaluation of the significance test usually the $t_{observed}$-value is calculated as shown in the slide and compared with the critical value from statistical tables for the required confidence level. For more details on significance testing see chapter 8 on 'Basic Statistics'.

If a Reference Material (RM), without a certified value or if an in-house material is used, there is no certified,

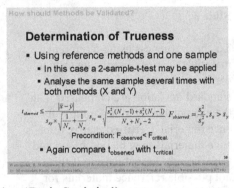

Determination of Trueness

- The formula can be re-arranged: $t_{observed} = \dfrac{|\bar{x} - x_{ref}|}{\dfrac{s}{\sqrt{n}}}$
- To be exact we also have to include the uncertainty of the reference value and get:

$$t_{observed} = \dfrac{\bar{x} - x_{ref}}{\sqrt{\left(\dfrac{s}{\sqrt{n}}\right)^2 + u_{ref}^2}}$$

- Compare $t_{observed}$ with $t_{critical}$ from statistical tables for the required level of confidence
- If $t_{observed} < t_{critical}$, there is no significant difference

traceable value. However, the procedure is the same. The value assigned to the material by the laboratory itself is the best available estimate for the 'true' value. Of course the determination of trueness is not as reliable as with a CRM.

Slide 39

To check trueness against an alternative method, the same sample or samples are analysed by the two methods and the two series of results are compared for a statistically significant difference. The samples may be CRM's, in-house standards, or simply typical samples. In the case where only one sample is used, the two methods are employed to make ten measurements.

A 2-sample-t-test is applied on the data (see 'Basic Statistics').

Determination of Trueness

- Using reference methods and one sample
  - In this case a 2-sample-t-test may be applied
  - Analyse the same sample several times with both methods (X and Y)

$$t_{observed} \leq \dfrac{|\bar{x} - \bar{y}|}{s_{xy} \times \sqrt{\dfrac{1}{N_x} + \dfrac{1}{N_y}}} \quad s_{xy} = \sqrt{\dfrac{s_x^2(N_x - 1) + s_y^2(N_y - 1)}{N_x + N_y - 2}} \quad F_{observed} = \dfrac{s_x^2}{s_y^2}, s_x > s_y$$

Precondition: $F_{observed} < F_{critical}$.

- Again compare $t_{observed}$ with $t_{critical}$

Slide 40

In case where many different samples are used, the samples are again analysed with the method under validation and with the reference method.

The data are used to perform a paired t-test (see 'Basic Statistics').

Determination of Trueness

- Using a reference method and many samples
  - Paired t-test
  - Analysed with our own method: $x_1, x_2, .., x_n$
  - Analysed with reference method: $c_{ref,1}, c_{ref,2}, ..., c_{ref,n}$

$$t_{observed} = \dfrac{|\bar{d}|}{\dfrac{s_d}{\sqrt{n}}}, \quad d = |x_i - c_{ref\,i}|$$

- Again compare $t_{observed}$ with $t_{critical}$

Slide 41

If no suitable reference materials are available, the determination of trueness can also be carried out via interlaboratory studies. The assigned value of the interlaboratory test sample, determined by the provider, is used as an estimate for the 'true' value. The organizers of such studies provide the participating laboratories with a report showing the score of each participant. From such scores the accuracy of the

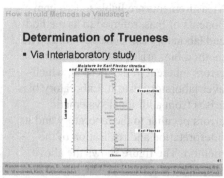

method used to analyse the circulated samples can be deducted. More details on this topic are discussed extensively in the separate chapter on 'Interlaboratory Tests'.

Slide 42

The laboratories can also determine the trueness of their methods using other in-house processes, for example, recovery studies of material, spiked with a known amount of the analyte. The recovery can be used as a measure of the trueness. Recoveries usually depend on sample matrix, sample preparation method and concentration present in the sample. The mean %

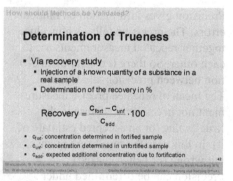

recovery for a trace component (< 0.1%) usually should be within 75 to 125 %. Care must be taken that the analyte that is spiked to the base material has a similar type of chemical binding form to the matrix as the original part of the analyte.

Recovery is defined as the fraction of analyte detected, compared with what was added to a test sample (fortified or spiked sample) prior to analysis.

Slide 43

Measurements are liable to two components of bias, referred to as method
and laboratory components of bias. The
method bias arises from systematic
errors inherent in the method, which-
ever laboratory uses it. Laboratory bias
arises from additional systematic
errors peculiar to the laboratory and its
interpretation and realisation of the
method. The figure shows the above
statement graphically. The bias of an
analytical method is usually determined by analysing relevant reference materials
or by recovery studies. The determination of overall bias with respect to appropriate
reference values is important in establishing traceability to recognized standards.

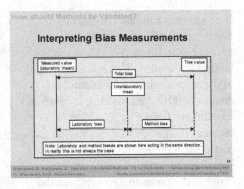

Slide 44

Precision gives information on random
errors. The precision tells us how close
together repeated measurements are to
each other. So there is a clear distinction between precision and trueness.
The mean of repeated precise measurement not necessarily is close to the
'true' value (or an accepted estimate of
it). In this case the results are precise,
but wrong (or better: biased).
Precision can be determined under
different conditions: repeatability, intermediate conditions (sometimes called
reproducibility-within-laboratory) and reproducibility.

Precision

Closeness of agreement between
indications or measured quantity values
obtained by replicate measurements on
the same or similar objects under
specified conditions [VIM]

Slide 45

Repeatability r, (the smallest expected precision) gives a measure of the variability to be expected when a method is performed by a single analyst on one piece of equipment over a short timescale, i.e. the variability to be expected when a sample is analysed in duplicate.

If a sample is analyzed by a number of laboratories for comparative purposes the precision measure used is reproducibility R, which is the largest measure of precision normally encountered. Reproducibility R is the precision taken under conditions where test results are obtained with the same method on identical test items in different laboratories with different operators using different equipment.

How should Methods be Validated?

Precision

- Precision under *repeatability* (r) conditions
  - Same method on identical test items, in the same laboratory, by the same operator, using the same equipment, within short time intervals
- Precision under *reproducibility* (R) conditions
  - Same method on identical test items, in different laboratories, with different operators, using different equipment

Slide 46

Intermediate precision includes within laboratories variations: different days, different analysts, different equipment, etc.

Intermediate precisions for in-between measures could also be considered to determine the precision between different analysts, over extended timescales etc. within a single laboratory.

The intermediate precision is what we encounter in the everyday laboratory practice. A laboratory can, for example, calculate the precision under repeatability conditions (same day, same method, same operator etc., but using different equipment, or same equipment but different operators etc.).

How should Methods be Validated?

Intermediate Precision R_w

- Repeatability conditions and reproducibility conditions are extreme cases
- Intermediate cases are most frequent
  - Same laboratory - different days
  - Same laboratory - different operators
  - Same laboratory - different equipment
  - etc.

The International Standard ISO 5725 establishes practical definitions of repeatability r and reproducibility R and provides basic principles for the layout, organization and analysis of precision experiments designed for estimating r and R.

Slide 47

Precision is usually expressed in terms of the standard deviation or relative standard deviation. The latter is useful in case of the comparison of standard deviation values that refer to the dispersion of individual results from samples with different concentrations. To calculate the standard deviations, the laboratory makes at least ten measurements (n=10) for each concentration level. The measurements are made under repeatability or reproducibility conditions or, more often, under intermediate precision conditions.

Evaluation of Precision

- Analyse the same sample several times under r, R, or R_w conditions
  - Standard Deviation

$$s = \sqrt{\frac{\sum_{i=1}^{n}(x_i - \bar{x})^2}{n-1}} \qquad RSD = \frac{s}{\bar{x}} \times 100$$

- The standard deviation can also be estimated from the average range of multiple measurements
  - For duplicate measurements the factor is 1.128 $s = \frac{\frac{1}{n}\sum(x_{max,i} - x_{min,i})}{1.128}$

Slide 48

The repeatability and reproducibility limits express the maximum acceptable difference between two consecutive measurements, which are done under repeatability or reproducibility conditions respectively. The factor $\sqrt{2}$ results from the error propagation of a difference. Each of the two values has a standard deviation of σ, so we get

$$r = t_\infty \cdot \sqrt{\sigma^2 + \sigma^2} = t_\infty \cdot \sqrt{2\sigma^2} = t_\infty \cdot \sqrt{2} \cdot \sigma$$

For 95% confidence t_∞ is 1.96.

Precision

- Repeatability limit r
  - With a specified confidence the difference between two values should be within that limit
  - For 95% confidence
  $r = t_{95\%} \cdot \sqrt{2} \cdot s_r$, for n → ∞ : r = 1.96 · $\sqrt{2} \cdot \sigma_r \approx 2.8 \cdot s_r$
- Reproducibility limit R
  - The same applies under reproducibility conditions
  $R = t_{95\%} \cdot \sqrt{2} \cdot s_R$, for n → ∞ : R = 1.96 · $\sqrt{2} \cdot \sigma_R \approx 2.8 \cdot s_R$
- With these formula r and R can be converted to s_r and s_R and vice versa

Slide 49

Sensitivity is the change in the response of a measuring instrument corresponding to the change of the concentration of the analyte. The sensitivity is therefore the slope of the response curve of the instrument. Where the response has been established as linear with respect to concentration i.e. within the linear range of

Sensitivity

- Change in the response of a measuring instrument divided by the corresponding change in the stimulus.
- The sensitivity A is defined as the slope of the calibration curve
  [IUPAC Orange Book]

the method, and the intercept of the response curve has been determined, sensitivity is a useful parameter for quantification.

Slide 50

The *robustness* of an analytical procedure is a measure of its capacity to remain unaffected by small, but deliberate variations in method parameters and provides an indication of its reliability during normal usage. The intra-laboratory study to examine the behaviour of an analytical process when small changes to the environment and/or operating conditions are made is called the *ruggedness test*.

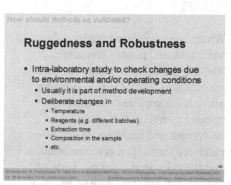

Such changes could be the temperature of the laboratory environment, the composition or pH of the mobile phase of an HPLC system, the suppliers of the reagents etc. Ruggedness is normally evaluated during method development, typically by the originating laboratory, before collaborating with other laboratories.

Slide 51

Most terms on this slide are explained in the chapters "Glossary" or "Basic Statistics" and thus are not repeated here.

Fortified materials/solutions: These are materials or solutions which have been fortified with the analyte(s) of interest. The fortification is usually made by spiking. These materials or solutions may already contain the analyte of interest. Note that most methods of fortification add the analyte in such a way that it will not be as closely bound to the sample matrix as it would be if it was present naturally.

Spiked materials: These are similar to fortified materials, indeed to some extent the terms are interchangeable. Spiking does not necessarily have to be restricted to the analyte of interest.

Incurred materials: These are materials in which the analyte of interest may be essentially alien, but has been introduced to the bulk at some point prior to the material being sampled. The analyte is thus more closely bound in the matrix than it would be had it been added by spiking. The following is one example of incurred material: Herbicides in flour from cereal sprayed with herbicides during its growth.

Slide 52

When using validated methods the analyst should always consider whether the existing validation data are adequate and, if yes, to consider his or her competence to use the method as well as the adequacy of equipment and facilities. When using validated methods it is recommended that the rules shown on this slide be followed in order to achieve an acceptable method performance. To re-evaluate is often called verification, i.e. to check if I can reach the specification as described in the validation documentation (e.g. in the standard).

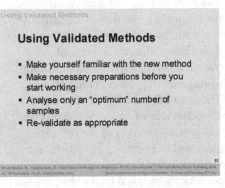

Slide 53

When using validated methods a number of questions must be answered:
- Are existent validation data adequate or is further validation required?
- Is the laboratory able to achieve the level of performance claimed in the method?
- Is the analyst sufficiently competent?
- Are the available equipment and facilities adequate?

Slide 54

QC or quality control describes the individual measures which relate to the monitoring and control of particular operations. QA or quality assurance relates to the overall measures taken by the laboratory to ensure and regulate quality.

Validation and Quality Control

- Internal QC

- External QC

Suitable quality control can be to analyse a known sample in each batch or after every five unknown samples. Internal QC implies in-house use of means mentioned on the next slide.

External QC can be external proficiency testing and comparison with other laboratories.

Slide 55

During the routine use of a method the samples used are not samples of known content which is often the case during the validation stage. To ensure that the performance of the method is similar to the performance during the validation stage, suitable control samples should be measured during routine operation. The quality control (QC) of the laboratory results can be either internal or external. The appropriate level of QC

Validation and Quality Control

- Internal QC includes
  - QC samples, use of control charts (recommended)
    - Warning and action limits
    - QC samples to be within limits
    - Realistic limits on the control chart
  - Various types of blanks to correct the response
  - Replicate analysis to check changes in precision
  - Blind analysis to check precision
  - Standards and chemical calibrants to check the stability of the response

depends on the reliability of the method, the criticality of the work etc. It is widely accepted that for routine analysis a level of internal QC of at least 5% (i.e. 1 control sample every 20 routine samples) is reasonable.

Internal QC includes the use of: blanks, chemical calibrants, spiked samples, blind samples, replicate analysis, and QC samples. QC samples should be homogeneous and stable. They should be available in sufficient quantities. The use of control charts is recommended (see chapter 13).

Slide 56

To monitor the laboratory performance against both its own requirements and the norms of peer laboratories there is a need for intercomparison of laboratory results. This is usually done by participating in proficiency testing (PT), i.e. an external quality assessment. In this way, the laboratory performance at national or international level are monitored and, furthermore, the reproducibility is highlighted.

Validation and Quality Control

- External QC
  - Proficiency testing
    - Monitoring of laboratory performance
    - Highlight reproducibility
    - Recognised by accreditation bodies
  - PT results as means of checking internal QC

Accreditation bodies recognize the benefit of these schemes and encourage laboratories to participate in PT. Regular participation in proficiency testing schemes is a recognised way to monitor laboratory performance (see chapter 15).

Slide 57

The method validation procedure should always be documented in order to give evidence of method performance. Proper documentation has an influence on the consistency and subsequent reproducibility, which, at the final stage affect the uncertainty contribution. The information in the validation documents should be clear and easily understood by everyone who uses the method.

Documentation of Validated Methods 1

- Documentation of the validation procedure
  - Clear and unambiguous implementation
  - Consistency during application
  - Large uncertainty contribution if inadequately documented methods
  - Information to be easily understood by everyone using the method

Slide 58

It can be a challenge to document a method properly. It can be a major challenge to work with a method where the reporter assumed a similar level of knowledge as his own. It is wise to give your write up to someone else and ask to work through it.

Documentation of Validated Methods 2

- When the validation process is complete document the procedures of the method (also important for auditing and evaluation purposes)
- With appropriate documentation
  - reuse of method is more consistent
  - uncertainty contribution is decreased
- Test quality of documentation with a competent colleague

Slide 59

A layout based on ISO 78-2:1999 is given in the EURACHEM guide "The Fitness for Purpose of Analytical Methods" under annex B. "House style" documentation is also suitable. It is wise however to have a single checking authority.

Documentation of Validated Methods 3

- Standard guidance for documentation is found in ISO 78 series
- "House style" is also adequate
- Documented methods form an important part of a lab's quality management system
  - when last updated

- Document control
  - is documentation complete
  - authorized for use
  - which version
  - which date
  - author
  - copyrights

Slide 60

ISO 78-2:1999 defines the layout needed for adequate documentation of chemical methods. The standard indicates a logical order for material with recommended headings and advice on the kind of information that should appear under each heading. Documented methods should be subject to an appropriate document control.

The Method Documentation Protocol

- Update & Review Summary
- Title
- Scope
- Warning & Safety precautions
- Definitions
- Principle
- Reagents & Materials
- Apparatus and equipment
- Sampling and samples

- Calibration
- Quality Control
- Procedure
- Calculation
- Reporting procedures including expression of results
- Normative references
- Appendix on method validation
- Appendix on measurement uncertainty

[Eurachem Guide Fitness for Purpose]

Slide 61

It is important that the analyst is able to translate validation data to give proper answers to customer's questions and offer solutions to their problems if and when required. The validation data support the validity of the results and enable the analyst to support his or her decisions. Therefore, access to the validation data is a must.

Special care is needed when dealing with issues such as method validation, variability and measurement uncertainty related to legal or forensic contexts. When reporting the results care must be taken to mention important details that indicate the quality of the results.

Interpretation of Validation Data

- Validation data to give answers and solutions to customer's problems
- Analyst's access to the validation data
- The analytical chemist as a technical advisor
  - Interpretation of the measurement uncertainty of results
  - Legal and forensic contexts
- When reporting the results
  - Either to correct for bias or to acknowledge bias
  - "Not detected" statement to be accompanied by the detection limit
  - Expression of uncertainties

Slide 62

Significant differences can easily be calculated, when precision data for repeatability and reproducibility are used. Quality controls based on the validation data can be used to confirm that the method is in control and producing meaningful results. Estimation of the measurement uncertainty enables expression of the result as a range of values in which the true value for the measurement can be said to lie within a stated level of confidence.

Implications of Validation Data for Calculating Results and Reporting

- Translate data in results to solve customers problem
- Consider:
  - Are observed differences significant?
    - precision data for repeatability and reproducibility
  - Quality controls confirm method is in control
    - based on validation data producing meaningful results
  - Estimation of measurement uncertainty enables
    - expression of results with a stated level of confidence

Slide 63

The customer may not need or understand the significance of the validation data. In such cases it can be safer to make the data available on request. However the analyst needs access to these data which can be used to support the validity of the results. Some people may not understand, why there is always an uncertainty attached to measurement. It is the analysts duty to treat those matter with care but openly.

Implications of Validation Data for Calculating Results and Reporting

- The analyst needs access to validation data to support validity of results
  - The customer may not need or understand it
- Information about uncertainties might not be understood by everybody
  - When required: uncertainty should be reported (confidence range e.g. 95 %)

Slide 64

The analytical methods should be used in the analytical laboratories as the tool to measure the parameters in a way that gives values that are as close to the true value as possible. The analytical methods should be validated and re-validated to a degree necessary to ensure that the results produced using such methods are sufficiently reliable. Validation data, if properly interpreted, can be used as a basis for comparison of the laboratory performance during routine analysis with the performance during the validation stage.

Summary

- The analytical method as a measurement tool
- Validation and verification
- Correct use of validation data
- Validation data as a measure of the method performance

Bibliography

AOAC - Terms and Definitions, available from www.aoac.org/terms.htm

Arvanitoyiannis I, Hadjicostas E (2001) Quality Assurance and Safety Guide for the Food and Drink Industry; CIHEAM/Mediterranean Agronomic Institute of Chania/European Commission MEDA

Australian Pesticides & Veterinary Medicines Authority (2004) Guidelines for the validation of analytical methods for active constituent, agricultural and veterinary chemical products, available from www.apvma.gov.au

Burgess C, Jones DG (1998) Equipment qualification for demonstrating the fitness for purpose of analytical instrumentation, Analyst 123, 1879-1886

Burns DT, Danzer K, Townshend A (2002) Use of the terms "Recovery" and "Apparent Recovery" in analytical procedures, Pure Appl. Chem. 74, 2201–2205

EURACHEM (1998) The Fitness for Purpose of Analytical Methods: A Laboratory Guide to Method Validation and Related Topics, available from www.eurachem.org

European Commission (2007) Method validation and quality control procedures for pesticide residues analysis in food and feed. SANCO/2007/3131, available from ec.europa.eu

Feinberg M, Laurentie M (2006) A global approach to method validation and measurement uncertainty, Accred Qual Assur 11, 3–9

Günzler H (1994): Accreditation and Quality Assurance in Analytical Chemistry, Springer, Berlin

Gustavo Gonzalez A, Angeles Herrador M (2007) A practical guide to analytical method validation, including measurement uncertainty and accuracy profiles. TrAC, Trends in Analytical Chemistry 26, 227-238

International Laboratory accreditation cooperation ILAC (2002) ILAC Policy on Traceability of Measurement Results, ILAC P-10, available from www.ilac.org

ISO 78-2:1999 Chemistry - Layouts for standards - Part 2: Methods of chemical analysis

ISO 78-3:1983 Chemistry - Layouts for standards - Part 3: Standard for molecular absorption spectrometry

ISO 78-4:1983 Chemistry - Layouts for standards - Part 4: Standard for atomic absorption spectrometric analysis

ISO 3534-1:2006 Statistics - Vocabulary and symbols - Part 1: General statistical terms and terms used in probability.

ISO/IEC Guide 99-12:2007 International Vocabulary of Metrology - Basic and General Concepts and Associated Terms, VIM, available from www.bipm.org.

IUPAC (1997) Orange book- "Compendium on Analytical Nomenclature", 3rd edition, available from www.iupac.org

MacNeil JD, Patterson J, Martz V (2000) Validation of analytical methods - proving your method is "fit for purpose", Special Publication - Royal Society of Chemistry

Muniz-Valencia1 R, Ceballos-Magana SG, Gonzalo-Lumbreras R, Santos-Montes A, Izquierdo-Hornillos R (2008) GC-MS method development and validation for anabolic steroids in feed samples, J. Sep. Sci. 31, 727–734

Neidhart B, Wegscheider W (2001): Quality in Chemical Measurements, Springer, Berlin

12   Measurement Uncertainty

Michael Koch

A measurement result is an information. Without knowledge of its uncertainty it is just a rumour.

A. Weckenmann

Measurement uncertainty is one of the key issues in quality assurance. It became increasingly important for analytical chemistry laboratories with the accreditation to ISO/IEC 17025.

The uncertainty of a measurement is the most important criterion for the decision whether a measurement result is fit for purpose. It also delivers help for the decision whether a specification limit is exceeded or not.

Estimation of measurement uncertainty often is not trivial. Several strategies have been developed for this purpose that will shortly be described in this chapter. In addition the different possibilities to take into account the uncertainty in compliance assessment are explained.

Slide 1

The uncertainty characterises the dispersion of the value, i.e. the expected spread of the data, which can be attributed to the measurand taking into account all reasonable effects.

Measurement Uncertainty

- "Non-negative parameter characterizing the dispersion of the quantity values being attributed to a measurand, based on the information used"
(VIM, 3rd edition)

Slide 2

"Uncertainty" is an unfortunate wording since for non-experts it implies doubts on the validity of results. Certainty would be a better choice, but it certainly will not be possible to change this term.
But on the contrary to the first impression the knowledge of the uncertainty of a result will increase the confidence in this value since it allows better judgement about the possible usage of this result.

Uncertainty – Doubt?

- Uncertainty of measurement does not imply doubt about the validity of a measurement
- On the contrary, knowledge of the uncertainty implies increased confidence in the validity of a measurement result.

Slide 3

Analytical measurement are never made without a reason. In most cases decisions are based on the results of measurement results and in many cases there is a lot of money behind these decisions. Therefore it often can save money to have an indication of the quality of these results.

Why is Measurement Uncertainty Important?

- Many important decisions are based on the results of chemical quantitative analysis
  - To estimate yields
  - To check materials against specifications or statutory limits
  - To estimate monetary value
  - ...
- Therefore it is important to have some indication of the quality of the results, i.e. the extent to which they can be relied on for the purpose in hand

Slide 4

If the customer is able to quantify his needs as a range for the accuracy of the results, than the measurement uncertainty is the adequate parameter for the comparison with these requirements.

Measurement Uncertainty and Fitness for Purpose

- Measurement uncertainty is the key figure for the decision whether an analytical result is fit for the customers purpose

Slide 5

One could think that the uncertainty of measurement is a perfect quality indicator for a laboratory. The lower the reported uncertainty the better the laboratory. As a consequence customers may commission these laboratories just because of the low reported uncertainties. It has to be clearly stated, that this is a misuse of uncertainty statements. As a result this would lead to unrealistic reported uncertainties by some laboratories just to get the contract. At the end of such an uncertainty-based battle all reported uncertainties are completely useless. Customers have to be educated to know that there is no need to have lower uncertainties than required from the purpose of the analysis.

Measurement Uncertainty – a Criterion for Placing Orders?

- Measurement uncertainties should not be misused as a quality indicator for laboratories
- Otherwise there will be a tendency among some laboratories to report unrealistic low measurement uncertainties
- There is no need to have lower uncertainties then requested from the customer

Slide 6

Measurement uncertainty especially is interesting when it comes to conformity assessment. In the graph the results of four different measurements are shown together with a specification limit. Without knowing the uncertainty of the measurements the situation seems to be very clear. Results 1 and 2 are below the specification, results 3 and 4 are above the limit.

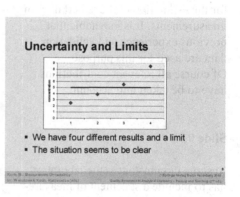

Uncertainty and Limits

- We have four different results and a limit
- The situation seems to be clear

Slide 7

Now the uncertainty is added to the results. For result 1 and 4 the situation is still the same. 1 is below, 4 is above the limit. But for the other results we are not sure any more. The uncertainty range tells us, that for both results there is a certain relevant probability that the value is above resp. below the limit.
We come back to this problem later (slide 53).

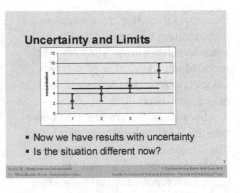

Slide 8

ISO/IEC 17025 also states requirements regarding measurement uncertainty. All laboratories must have procedures for the estimation of the uncertainty of measurements. It is inevitable that previous experience and validation data are used for this purpose.
Of course all relevant contributions have to be taken into account.

Measurement Uncertainty in ISO/IEC 17025

- 5.4.6.2 Testing laboratories shall have and shall apply procedures for estimating uncertainty of measurement
- ... Reasonable estimation shall be based on knowledge of the performance of the method and on the measurement scope and shall make use of, for example, previous experience and validation data
- 5.4.6.3 When estimating the uncertainty of measurement, all uncertainty components which are of importance in the given situation shall be taken into account using appropriate methods of analysis

Slide 9

In some circumstances the laboratory has to include a statement of measurement uncertainty also in its report, especially when the measurement uncertainty is relevant for the interpretation of the data with regard to a specification limit (see slides 53).

ISO/IEC 17025 - Report of Measurement Uncertainty

- Test reports, 5.10.3.1:
  ... tests reports shall, where necessary for the interpretation of the test results, include the following:
  ...
  c) where applicable, a statement on the estimated uncertainty of measurement; information on uncertainty is needed in test reports when it is relevant to the validity or application of the test results, when a client's instruction so requires, or when the uncertainty affects compliance to a specification limit;

Slide 10

Measurement uncertainty is the key figure for the decision if a measurement result is fit for the customer's purpose with regard to the accuracy of the measurement. It is part of the validation of the method to find out if these requirements are met. In many cases the problem is that the customers are not used to specify these requirements. Then the laboratory has to clarify this in dialogue with its customer.

Customer Needs

- Before calculating or estimating the measurement uncertainty, it is recommended to find out what are the needs of the customers
- The main aim of the actual uncertainty calculations will be to find out if the laboratory can fulfil the customer demands
- Customers often are not used to specify demands, so in many cases the demands have to be set in dialogue with the customer.

Slide 11

The "Guide to the Expression of Uncertainty in Measurement" (GUM) to some extent is the "Bible" of uncertainty estimation. Since it was originally made for physical measurements in metrology laboratories, it is quite difficult to translate it into analytical chemistry problems, especially for routine measurements.

Basic Document

- "Guide to the Expression of Uncertainty in Measurement", published 1993 by ISO in collaboration with BIPM, IEC, IFCC, IUPAC, IUPAP and OIML.
- Comprehensive document for all kind of measurements
- Difficult to apply for many chemists

Slide 12

To help other chemists the colleagues from an EURACHEM/CITAC working group did that transformation and fortunately included some examples. The guide is available from the internet free of charge.

Interpretation of GUM for Chemists

- EURACHEM/CITAC-Guide "Quantifying Uncertainty in Analytical Measurement"
- 2nd edition published in 2000
- Available from www.eurachem.org

Slide 13

The measurement uncertainty of the final result depends on many different contributions (uncertainty sources). The listing in the slide shows some of them, but does not claim complete-ness.

Uncertainty Sources

- In practice the uncertainty on the result may arise from many possible sources, including examples such as
  - Incomplete definition,
  - Sampling,
  - Matrix effects and interferences,
  - Environmental conditions,
  - Uncertainties of masses and volumetric equipment,
  - Reference values,
  - Approximations and assumptions incorporated in the measurement method and procedure, and
  - Random variation

Slide 14

Since the measurement uncertainty expresses a range within which we expect the true value with a certain probability, we always have to state this probability (the level of confidence). If we express it as a standard deviation (i.e. with a probability of approx. 68%) we call it standard uncertainty.

Standard Uncertainty

- When expressed as a standard deviation, an uncertainty component is known as a standard uncertainty u

Slide 15

If we combine some standard uncer-tainty contributions from different sources then we calculate that accord-ing to the law of propagation of errors (square root of the sum of squares) and call it combined (standard) uncertainty.

Combined Standard Uncertainty

- For a measurement result y
- The total uncertainty,
  - Termed combined standard uncertainty and
  - Denoted by $u_c(y)$,
- Is an estimated standard deviation
- Calculated according to the law of error propagation
  - From the positive square root of the total variance obtained by combining all the uncertainty components, however evaluated

$$u_c(y) = \sqrt{u_1(y)^2 + u_2(y)^2 + ... + u_n(y)^2}$$

Slide 16

In most cases a confidence level of 68% is not enough to take reliable decisions. To increase this confidence the combined standard uncertainty is multiplied with a factor to get an expanded uncertainty. If we choose a factor of 2 we get a level of confidence of approx. 95%. This is the most widely used expanded uncertainty in analytical chemistry.

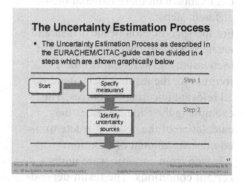

Slide 17

The process to estimate the uncertainty as described in the GUM as well as in the EURACHEM/CITAC guide is divided in four steps.
The first is the clear and unambiguous specification what has to be measured under which conditions. This some-times is more tricky than it seems to be, since it is very much connected to the second step, the identification of uncertainty sources. These sources also include parameters that do not directly go into the calculation of the result, but nevertheless influence the result and therefore the uncertainty.

Slide 18

If we have a complete list of sources we try to group those that are covered by existing data (e.g. from repeatabil-ity experiments). The next step is the quantification of the grouped compo-nents. Usually some components will remain that have to be quantified separately. Step 3 is finalised with the conversion of all uncertainty compo-nents into standard uncertainties.

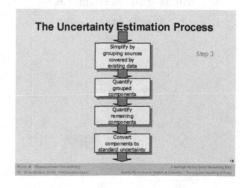

Slide 19

The last step is the combination of all sources and the review of the largest contribution. Perhaps it will be possible to get a better estimation of these components that influence the magnitude of the combined uncertainty most. Finally the expanded uncertainty is calculated to adjust it to the requested level of confidence.

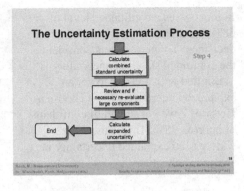

Slide 20

Let us look a bit more in detail on the different steps:

For the specification of the measurand we need a statement of what we want to measure and at the same time a formula for the result which contains all relevant uncertainty sources. The example in the slide describes the calculation of the result of a determination of the amount of cadmium released from ceramic ware under certain conditions. The result depends on the content of Cd in the extraction solution c_0, the volume of the leachate V_L, the surface area a_V that is extracted and possibly a dilution factor. These parameters are used to calculate the result. But we also have to consider that the acid concentration, the extraction time and the temperature are influencing the result. Since they are not directly involved in the calculation of the result, we add factors with the value 1. But we assume that this value 1 will have an uncertainty as well.

Slide 21

In the 2nd step we try to figure out all relevant uncertainty sources that influence the parameters identified in step 1. The figure shows a fishbone or *Ishikawa* diagram that is helpful to get an overview.

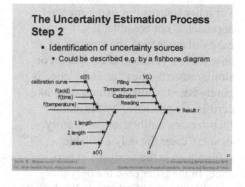

Slide 22

When we quantify the uncertainty sources or groups of sources we have to consider that we can assume some to be normally distributed. In this case we get the standard uncertainty directly from the standard deviation. In other cases we may only have the information that the value is somewhere between two limits. If we don't have information about the type of distribution we assume a rectangular distribution, where all values between the limits have the same probability. The standard deviation, and therefore our standard uncertainty, then is calculated as

$$s = \sqrt{a^2/3}$$

with *a* being the halfwidth of the rectangular distribution.

Slide 23

Uncertainties from some sources may be quantified by doing experiments. From the results of repeated measurements we get a standard deviation which we can use directly as an estimate for the standard uncertainty. This is called type A evaluation of uncertainties.

Slide 24

For other sources this experimental approach is not possible. E.g. for the purity of a material, the trueness of a balance or a volumetric equipment or similar uncertainties our knowledge is purely based on other information like certificates published by the manufacturer or a calibration laboratory. In some cases the only available information is the analyst's experience. In the latter case the only thing possible is a qualified guess. We have to find the best way how to convert the available knowledge into a standard uncertainty. All these types of estimation are called type B evaluation.

The Uncertainty Estimation Process Step 3
- Type B evaluation of uncertainties
  - Other components are evaluated from assumed probability distributions based on experience or other information (e.g. a rectangular distribution of the purity of a chemical where no other information is given than >99%
  - They also can be characterised by standard deviations

Slide 25

From the above mentioned estimation process we get the standard uncertainty of the parameter influencing the result. But we are interested in the component of the uncertainty of the *result* caused by that parameter. So we have to convert the first one into the latter one. This is done by multiplying the uncertainty of the parameter with a sensitivity coefficient which can be obtained from the partial derivation of the result with respect to the respective parameter.

The Uncertainty Estimation Process Step 4
- Conversion of the uncertainty of a parameter into an uncertainty component of the result
  - The extend to which the uncertainty of the result is influenced by this parameter is reflected by the "quantitative expression" determined in step 1.
  - The related sensitivity coefficient is determined by partial derivation of y with respect to our parameter x_i

$$u(y, x_i) = c_i \cdot u(x_i) = \frac{\partial y}{\partial x_i} \cdot u(x_i)$$

  - $u(y,x_i)$ denotes the uncertainty in y arising from the uncertainty in x_i

Slide 26

The combined uncertainty we get from the square root of the sum of squares of all uncertainty contributions.

The Uncertainty Estimation Process Step 4

- Calculation of Combined Uncertainty
  - The general relationship between the combined standard uncertainty $u_c(y)$ of a value y and the uncertainty arising from the independent parameters $x_1, x_2, ...x_n$ on which it depends is according to the law of propagation of errors

$$u_c(y(x_1, x_2,...)) = \sqrt{\sum_{i=1}^{n} u(y, x_i)^2}$$

Slide 27

The last step is the calculation of the uncertainty with a higher level of confidence, the expanded uncertainty. As described above this is done by multiplying with a coverage factor k.

The Uncertainty Estimation Process Step 4

- Calculation of Expanded Uncertainty
  - The combined uncertainty is multiplied with the coverage factor

$$U = k \cdot u_c$$

  - A coverage factor of 2 delivers a confidence level of about 95%

Slide 28

As already mentioned the modelling approach described in the GUM is difficult to translate into analytical chemistry because the testing procedures are very complex and it is a huge effort to identify and quantify all uncertainty sources. For routine laboratories often handling dozens or even hundreds of different test methods it is nearly impossible to cope that with the modelling approach.

Problems in Chemical Laboratories

- Chemical analyses often are very complex testing procedures with lots of uncertainty sources that can hardly be quantified separately
- The calculation of a complete uncertainty budget for all routine tests often is impossible for chemical laboratories

Slide 29

Having this in mind other approaches were developed that use already existing quality control and validation data. One of the most widely known publication for that is the NORDTEST "Handbook for calculation of measurement uncertainty in environmental laboratories".

NORDTEST Handbook

- NORDTEST, a cooperation of laboratories in the Nordic countries, published a „Handbook for calculation of measurement uncertainty in environmental laboratories"
- To provide a practical, understandable and common way of measurement uncertainty calculations, mainly based on already existing quality control and validation data covering all uncertainty sources in a integral way
- Available as TR 537 from: www.nordicinnovation.net/nordtest.cfm

Slide 30

The Handbook describes two possibilities where one is using quality control and validation data with the aim to come up with data for the imprecision (lack of precision) and for bias (lack of trueness).
The other method uses data from inter-laboratory tests directly.

NORDTEST-Approach – 2 possibilities

a) Combination of
  - Reproducibility within the laboratory and
  - Estimation of the method and laboratory bias

b) Use of reproducibility between laboratories more or less directly (as described also in ISO/TS 21748:2004)

Slide 31

This flowchart is describing the whole process divided in 6 steps:
1. Specification of the measurand (as described above)
2. Quantification of the within laboratory reproducibility (or better the imprecision)
3. Quantification of the laboratory and method bias
4. Conversion into standard uncertainties
5. Calculation of the combined uncertainty
6. Calculation of the expanded uncertainty

Flowchart for Method a)

Specify measurand

Quantify components for within lab reproducibility
A control samples
B possible steps, not covered by the control sample

Quantify bias components

Convert components to standard uncertainty

Calculate combined standard uncertainty

Calculate expanded uncertainty $U = 2 \cdot u_c$

Slide 32

We are starting with the case where we have a control sample that covers the whole analytical process including all sample preparation steps. The matrix of the control sample is similar to that of the routine samples. Then the standard deviation of the analysis of this sample (under between-batch conditions) can be used directly as an estimate for the reproducibility within the laboratory. The standard deviation

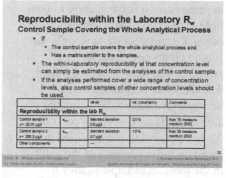

can be taken directly from a control chart for this control sample (see chapter13). In the table two examples are shown for different concentration levels.

Slide 33

If we don't have such an ideal control sample, but only one with a matrix different from the routine sample (e.g. a standard solution) than we have to consider also the uncertainty component arising from changes in the matrix. For this purpose we use the (repeatability) standard deviation calculated from repeated measurements of our routine samples (performed e.g. for a range control

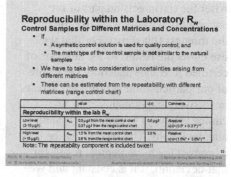

chart). When we estimate the reproducibility within laboratory we now have to combine both contributions by calculating the square root of the sum of squares. In the table we have again two examples, one using absolute values (because we assume the absolute uncertainty to be nearly constant in this concentration range) and one using relative values (because we assume that the relative uncertainty is more constant for higher concentrations). The repeatability of the analysis is included in both components. We accept this (usually small) incorrectness in order to keep it simple.

Slide 34

In some cases it is completely impossible to prepare control samples for the use in X-charts. So we have only the second component from the last example. The long-term component that takes into account changes between series has to be estimated differently. If we don't have other data, the last possibility is a qualified guess based on the experience of the analyst. In the table an example is shown

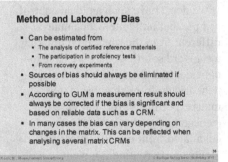

where the analyst expects the between-series standard deviation in the range of about 0.5% based on observations during the calibration of the oxygen sensor.

Slide 35

After the (im)precision part of the uncertainty we now want to estimate the trueness (bias) part. For that we use either the results of the analysis of certified reference materials, the results of our participation in proficiency tests or recovery experiments. The first choice is always to eliminate systematic deviations. The second choice is, to correct the measurement result for that deviation, if it is constant and we have reliable information on it. If both are not possible we include the bias component in our measurement uncertainty estimate.

Slide 36

The bias part of the uncertainty also has two parts: the bias itself and the uncertainty of the nominal value. If only one certified reference material is available we have to include the uncertainty of the determination of the bias in addition. This is done by using the standard deviation of this bias determination. All components are combined as shown in the slide.

Method and Laboratory Bias u_{bias}
Components of Uncertainty

- The bias (as % difference from the nominal or certified value)
- The uncertainty of the nominal/certified value $u(C_{ref})$ u_{bias} can be estimated by:

$$u_{bias} = \sqrt{RMS_{bias}^2 + u(C_{ref})^2} \quad \text{with} \quad RMS_{bias} = \sqrt{\frac{\sum(bias_i)^2}{n}}$$

- And if only one CRM is used

$$u_{bias} = \sqrt{(bias)^2 + \left(\frac{s_{bias}}{\sqrt{n}}\right)^2 + u(C_{ref})^2}$$

Slide 37

We start with an example where only one certified reference material is available with a certified value of 11.5 and an expanded uncertainty (95%) of 0.5. First we convert the expanded uncertainty into a standard uncertainty by dividing it by 1.96. Now we have the standard uncertainty of the reference value.

Method and Laboratory Bias u_{bias}
Use of One Certified Reference Material

- The reference material should be analysed in at least 5 different analytical series
- Example: Certified value: 11.5 ± 0.5 (95% confidence interval)

Uncertainty component from the uncertainty of the certified value	
Convert the confidence interval	The confidence interval is ± 0.5. Divide this by 1.96 to convert it to standard uncertainty: 0.5/1.96=0.26
Convert to relative uncertainty $u(C_{ref})$	100·(0.26/11.5)=2.21 %

Slide 38

The next step is the quantification of the bias itself from the mean of the results of analysis of the CRM and the standard deviation of these measurements. Using the formula from slide 36 and the uncertainty of the reference value from the previous slide we get the bias uncertainty component.

Method and Laboratory Bias u_{bias}
Use of One Certified Reference Material

- Quantify the bias
  - The CRM was analysed 12 times. The mean is 11.9 with a standard deviation of 2.2%
  - This results in:
    bias = 100 · (11.9 – 11.5) / 11.5 = 3.48% and
    s_{bias} = 2.2% with n = 12
  - Therefore the standard uncertainty is:

$$u_{bias} = \sqrt{(bias)^2 + \left(\frac{s_{bias}}{\sqrt{n}}\right)^2 + u(C_{ref})^2} =$$

$$\sqrt{(3.48\%)^2 + \left(\frac{2.2\%}{\sqrt{12}}\right)^2 + 2.21\%^2} = 4.2\%$$

Slide 39

If we are lucky and we have several CRMs available, the uncertainty of the bias determination is covered by the results of the measurement of the different materials. From the biases of each CRM analysis we calculate the root of the mean of squares as shown in the slide. Finally we combine this value with an average of the uncertainty of the certified value and we get the final uncertainty component.

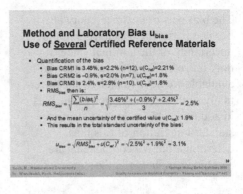

Method and Laboratory Bias u_{bias}
Use of <u>Several</u> Certified Reference Materials

- Quantification of the bias
  - Bias CRM1 is 3.48%, s=2.2% (n=12), $u(C_{ref})$=2.21%
  - Bias CRM2 is –0.9%, s=2.0% (n=7), $u(C_{ref})$=1.8%
  - Bias CRM3 is 2.4%, s=2.8% (n=10), $u(C_{ref})$=1.8%
  - RMS_{bias} then is:

$$RMS_{bias} = \sqrt{\frac{\sum(bias)^2}{n}} = \sqrt{\frac{3.48\%^2 +(-0.9\%)^2 +2.4\%^2}{3}} = 2.5\%$$

  - And the mean uncertainty of the certified value $u(C_{ref})$: 1.9%
  - This results in the total standard uncertainty of the bias:

$$u_{bias} = \sqrt{RMS_{bias}^2 + u(C_{ref})^2} = \sqrt{2.5\%^2 +1.9\%^2} = 3.1\%$$

Slide 40

If no CRM is available at all, but we participated in a proficiency testing scheme, we can use these results also for the estimation of the bias uncertainty component, provided that we assume the assigned value in the PT to be a good estimate for the true value. Of course one PT participation is not enough. At least six PT participations are recommended. Again we have to quantify the uncertainty of the nominal

Method and Laboratory Bias u_{bias}
Use of PT Results

- In order to have a reasonably clear picture of the bias from interlaboratory comparison results, a laboratory should participate at least 6 times within a reasonable time interval

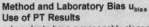

Uncertainty component from the uncertainty of the nominal value	
Between laboratory standard deviations s_R	s_R has been on average 9% in the 6 exercises
Convert to relative uncertainty $u(C_{ref})$	Mean number of participants= 12 $u(C_{ref}) = \frac{s_R}{\sqrt{n}} = \frac{9\%}{\sqrt{12}} = 2.6\%$

value (i.e. the assigned value in the PT) and the bias itself. Sometimes the uncertainty of the assigned value is delivered by the PT provider. If not, the uncertainty of a consensus mean can be estimated from the reproducibility standard deviation and the number of participants as shown in the slide.

Slide 41

For the calculation of the bias itself we again use the root of the mean of squares of all biases. In the example shown we have 6 PT results. We calculate the relative bias of these values and then the RMS_{bias}. Finally we combine the RMS_{bias} with the uncertainty of the assigned value and we get the uncertainty component for the bias.

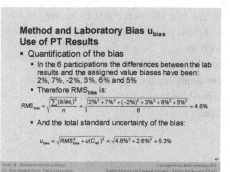

Slide 42

A third possibility to estimate the bias component of the uncertainty is to perform a recovery experiment, where a sample is spiked with the analyte. The uncertainty of the nominal value (i.e. the spiked amount) has to be calculated from the uncertainty of the volume added and the concentration of the spike solution. In the example in the slide the spiking is done using a micropipette. The uncertainty of the concentration of the spike solution is taken from the certificate of the manufacturer of the standard solution used and the uncertainty (repeatability and max. bias) of the micropipette is delivered by the respective manufacturer. Both components are combined to the uncertainty of the spike.

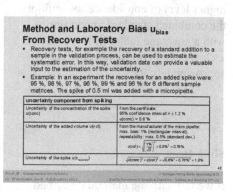

Slide 43

Now we quantify the bias itself in the same way as above. We calculate the RMS_{bias} of the bias (100% - recovery). Finally we combine the two contribution and we get u_{bias}.

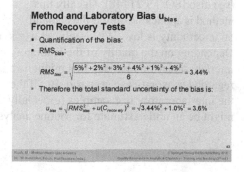

Slide 44

In the last slides we got to know different possibilities to estimate R_w as well as u_{bias}. To get the overall combined standard uncertainty u_c of our analyses we again combine these contributions as root of the sum of squares.

Combination of the Uncertainties
(Reproducibility within the Laboratory and Bias)

- Reproducibility (R_w) (from control samples and other estimations)
- Bias u_{bias} (from CRM, PT or recovery tests)
- Combination:

$$u_c = \sqrt{u(R_w)^2 + u_{bias}^2}$$

Slide 45

To get an expanded uncertainty with a higher level of confidence we multiply the combined standard uncertainty with a coverage factor k. If we choose k=2, we get a confidence level of approximately 95%.

Calculation of the Expanded Uncertainty

- For the conversion to an approx. 95% confidence level

$$U = 2 \cdot u_c$$

Slide 46

If we fail with the method described above because of lack of data or for other reasons, we also have the possibility to use the reproducibility standard deviation taken from an inter-laboratory comparison, which was either performed for the validation of the method (e.g. during standardization) or for proficiency testing of laboratories (see also ISO/TS 21748). Usually this method is applicable only if the demand

Method b) - Direct Use of Reproducibility Standard Deviations

- If the demand on uncertainty is low
- $u_c = s_R$
- The expanded uncertainty becomes $U = 2 \cdot s_R$
- This may be an overestimate depending on the quality of the laboratory – worst-case scenario
- It may also be an underestimate due to sample inhomogeneity or matrix variations

on uncertainty is low. We simply set $u_c = s_R$ and therefore we get $U = 2 \cdot s_R$. Depending on the quality of the laboratory compared with those participating in interlaboratory comparison this might be an overestimate and therefore can be regarded as worst-case scenario. But if the routine samples are more complex than the interlaboratory test samples (especially with regard to inhomogeneity) than it might be an underestimate for routine analysis.

Slide 47

If we take the reproducibility standard deviation from a standard we have to prove that we are able to perform in accordance with the standard, i.e. that we have no reasonable bias and the repeatability of our measurement is close to that described in the standard. If this is the case, our expanded uncertainty U is twice the reproducibility standard deviation.

Reproducibility Standard Deviation from a Standard

- The laboratory must first prove that they are able to perform in accordance with the standard method
  - No "unusual" bias
  - Verification of the repeatability
- The expanded uncertainty then is:

$$U = 2 \cdot s_R$$

Slide 48

This slide shows an example taken from an European standard for the analysis of mercury in water. For drinking water the standard states a reproducibility variation coefficient of 30% on a mercury concentration level of 0.8 µg/l. Provided that we can prove that we can perform as described in the standard our expanded measurement uncertainty (95% confidence) is estimated to 60%.

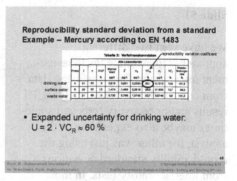

Reproducibility standard deviation from a standard
Example – Mercury according to EN 1483

- Expanded uncertainty for drinking water:
  U = 2 · VC$_R$ ≈ 60 %

Slide 49

We also can take suitable data from a proficiency test (PT). In this case the laboratory must have participated in the PT successfully. We also have to consider, if the PT covered all relevant uncertainty components and steps of analysis. This includes e.g. if the matrix of the PT sample was similar to routine samples. If this is the case, we again calculate U from 2 · s$_R$.

Reproducibility Standard Deviation from a PT

- The laboratory must have been successfully participating in the PT
- If the comparison covers all relevant uncertainty components and steps (matrix?)
- The expanded uncertainty then also is:

$$U = 2 \cdot s_R$$

Slide 50

This slide shows an example from a drinking water PT provided by the University of Stuttgart. At the same level as in the previous example we find a reproducibility variation coefficient of about 20%. So our expanded uncertainty is estimated to 40%. Without doing anything, but looking into some papers, we therewith have two estimates of our expanded uncertainty, 60% and 40%. This shows the uncertainty of this uncertainty estimation.

Reproducibility Standard Deviation from a PT Example – Mercury in a Univ. Stuttgart PT

- $u_c = s_R \approx 20\%$
- $U \approx 40\%$

Slide 51

We have seen two different approaches to estimate the measurement uncertainty. One was using data from control charts, CRM analysis, PT results and/or recovery tests and sometimes maybe also experience of the analyst, the other was just using the reproducibility standard deviations from interlaboratory tests. In most cases the second method delivers higher estimates.

Summary of the NORDTEST Approach

- Two different methods to estimate the measurement uncertainty have been introduced:
- Method a)
  - Estimation of the within-lab reproducibility (mainly from control charts)
  - Estimation of the bias (from analyses of CRM, PT results or recovery tests)
  - Combination of both components
- Method b)
  - direct use of the reproducibility standard deviation from standards or PTs as combined standard uncertainty
- As a rule method b) delivers higher measurement uncertainties (conservative estimation)

Slide 52

If we report the measurement uncertainty to our customer we always have to state also the level of confidence of the range we report as uncertainty. Otherwise the uncertainty statement is completely useless. It could be helpful, also to state the method that was used to estimate the uncertainty.

Expression of Uncertainty

- The statement of uncertainty always has to contain the level of confidence
- If possible also state the estimation method used
- Example:

  SO_4^{2-} in waste water (ISO 10304-2): 100 ± 8 mg/l*

  * Measurement uncertainty was derived from results of interlaboratory comparisons. It represents an expanded uncertainty with a coverage factor k=2; this corresponds to a level of confidence of about 95%.

Slide 53

Once we know an estimate of the measurement uncertainty another problem comes up. How do we handle those uncertainties in compliance assessment? The ideas described here are taken from a EURACHEM/CITAC- Guide.

Uncertainties and Limits

- How can we handle uncertainties in the assessment of values with respect to limits
- From: EURACHEM/CITAC Guide "Use of uncertainty information in compliance assessment, 2007

Slide 54

Let us assume we have an upper limit for compliance of a product and we have four different values, where one is together with its complete expanded uncertainty range above the limit (i) and one is completely below the limit (iv). The other two values have ranges of expanded uncertainty that include the limit. In one case the measurement result itself is above the limit (ii), in the other case below(iii).

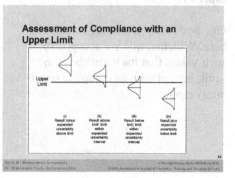

Assessment of Compliance with an Upper Limit

Now we have to decide, which of the four tested products are compliant and which are not.

Slide 55

The key for that decision is a decision rule. The most important point is to decide about an acceptable level of probability of making a wrong decision and who has to take this risk. The next slides will explain how such decision rules can be developed.

Decision Rules

- The key to the assessment are the „decision rules"
- Based on
  - The measurement result,
  - Its uncertainty,
  - The specification limit
- And taking into account
  - The acceptable level of the probability of making a wrong decision
- These rules give a prescription for the acceptance or rejection of a product

Slide 56

We have to formulate decision rules in such a way that we define acceptance zones (where the product is declared compliant) and rejection zones (where the product is declared non-compliant).

Acceptance Zone – Rejection Zone

- Based on the decision rules such zones are defined
  - If the measurement lies in the acceptance zone the product is declared compliant
  - If the measurement lies in the rejection zone the product is declared non-compliant

Slide 57

One possibility is the rule described here. With that rule we accept also such values that itself are above the limit, but not with its complete uncertainty range.

Simple Decision Rules – 1st Example

- „A result implies non-compliance with an upper limit if the measured value exceeds the limit by the expanded uncertainty."
  - With this decision rule only case (i) in the figure would imply non-compliance

Slide 58

This is a rule that works the other way around. Compliance is stated only when the value *and* its complete uncertainty range is below the limit.

Simple Decision Rules – 2nd Example

- „A result implies non-compliance with an upper limit if the measured value exceeds the limit minus the expanded uncertainty."
  - With this decision rule only case (iv) in the figure would imply compliance

Slide 59

A widely used example is not to take into account the uncertainty when deciding upon compliance. But in this case it necessary to specify a maximum allowed uncertainty. Otherwise the risk of making a wrong decision is unknown.

**Simple Decision Rules –
Another Widely Used**

- „A result equal to or above the upper limit implies non-compliance and a result below the limit implies compliance, provided that the uncertainty is below a specified value."
  - This is normally used where the uncertainty is so small compared with the limit that the risk of making a wrong decision is acceptable
  - To use such a rule without specifying the maximum permitted value of the uncertainty would mean that the probability of making a wrong decision would not be known
  - With this decision rule, cases (i) and (ii) would imply non-compliance, cases (iii) and (iv) compliance

Slide 60

Of course it is also possible to decide that in cases where the limit is within the uncertainty range, other measures have to be taken, e.g. more measurements (to reduce the uncertainty range) or to decide to sell the product at a different price.

**More Complicated
Decision Rules**

- Decision rules may include, for example, that for cases (ii) and (iii) in the figure, additional measurement(s) should be made, or
- That the manufactured product might be compared with an alternative specification to decide on possible sale at a different price

Slide 61

To define the above mentioned acceptance and rejection zones we need a specification and a decision rule.

**Basic Requirements for the
Decision**

- A **specification** giving upper and/or lower permitted limits of the characteristics (measurands) being controlled
- A **decision rule** that describes how the measurement uncertainty will be taken into account with regard to accepting or rejecting a product according to its specification and the result of a measurement
- The limit(s) of the **acceptance or rejection zone** (i.e. the range of results), derived from the decision rule, which leads to acceptance or rejection when the measurement result is within the appropriate zone

Slide 62

If we use the decision rule from the 1st example (slide 57) the acceptance zone increases the upper limit. That means that we have a high confidence of correct rejection. If we transform this to the sale of a commercial product, the purchaser takes the risk of a wrong decision.

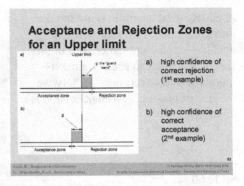

If we use the decision rule from the 2nd example (slide 58), the seller takes the risk of a wrong decision.

The decision that does not take into account the uncertainty, does not create any "guard bands", i.e. seller and purchaser are sharing the risk. But in order to 'know' what risk they are sharing, they should know about a maximum uncertainty.

Slide 63

Of course it is possible to define decision rules also in cases where we have upper and lower limits. The example here shows a case where the risk of false acceptance is low, the seller has to take the risk of a wrong decision. A wrong decision would mean that the seller is not able to sell a product which is in fact compliant.

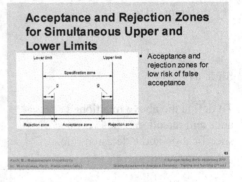

Slide 64

In some fields where measurement results are used, it is not yet common practice to explicitly define decision rules. The best way to do it, is in the product specification or the regulation. Where this is not the case, it should be clarified as part of the analytical requirement by the customer. When compliance or non-compliance is reported, it should be clear to all involved parties, which decision rules were used.

Who Defines the Decision Rule?

- The relevant product specification or regulation should ideally contain the decision rules
- Where this is not the case then they should be drawn up as part of the definition of the analytical requirement (i.e. during contract review)
- When reporting on compliance, the decision rules that were used should always be made clear

Slide 65

A lot of guidance on how to estimate measurement uncertainties is available. All guides are freely available from the internet.

Guidance Documents

- ISO Guide 98: "Guide to the expression of uncertainty in measurement" (www.bipm.org)
- EURACHEM/CITAC: Quantifying Uncertainty in Analytical Measurement, 2nd Edition (2000) (www.eurachem.org)
- NORDTEST: Handbook for calculation of measurement uncertainty in environmental laboratories. Report TR 537 (www.nordicinnovation.net/nordtest.cfm)
- LGC/VAM: Development and Harmonisation of Measurement Uncertainty Principles Part(d): Protocol for uncertainty evaluation from validation data (www.vam.org.uk)
- Niemela, S.I.: Uncertainty of quantitative determinations derived by cultivation of microorganisms. MIKES-Publication J4/2003 (www.mikes.fi)
- EA Guidelines on the Expression of Uncertainty in Quantitative Testing EA-4/16 (rev.00) 2003 (www.european-accreditation.org)
- ILAC-G17:2002 Introducing the Concept of Uncertainty of Measurement in Testing in Association with the Application of the Standard ISO/IEC 17025 (www.ilac.org)
- A2LA: Guide for the Estimation of Measurement Uncertainty In Testing, 2002 (www.a2la.net)
- EUROLAB: Measurement uncertainty revisited: Alternative approaches to uncertainty evaluation, 2007 (www.eurolab.org)
- EURACHEM/CITAC Guide "Use of uncertainty information in compliance assessment, 2007 (www.eurachem.org)

Bibliography

A2LA (2002) Guide for the Estimation of Measurement Uncertainty In Testing, available from www.a2la.net

EA (2003) Guidelines on the Expression of Uncertainty in Quantitative Testing EA-4/16 (rev.00), available from www.european-accreditation.org

EURACHEM/CITAC (2000) Quantifying Uncertainty in Analytical Measurement, 2nd Edition, available from www.eurachem.org

EURACHEM/CITAC (2007) Use of uncertainty information in compliance assessment, available from www.eurachem.org

EUROLAB (2007) Measurement uncertainty revisited: Alternative approaches to uncertainty evaluation, available from www.eurolab.org

ILAC (2002) G17 - Introducing the Concept of Uncertainty of Measurement in Testing in Association with the Application of the Standard ISO/IEC 17025, available from www.ilac.org

ISO/IEC Guide 98:1995 Guide to the expression of uncertainty in measurement (GUM), also available as JCGM 100:2008 from www.bipm.org

ISO/IEC Guide 99:2007 International Vocabulary of Metrology – Basic and General Concepts and Associated Terms (VIM), 3rd edition, also available as JCGM 200:2008 from www.bipm.org

ISO/TS 21748:2004 Guidance for the use of repeatability, reproducibility and trueness estimates in measurement uncertainty estimation

LGC/VAM (2000) Development and Harmonisation of Measurement Uncertainty Principles Part(d): Protocol for uncertainty evaluation from validation data, available from www.vam.org.uk

Niemelä SI (2003) Uncertainty of quantitative determinations derived by cultivation of microorganisms. MIKES-Publication J4/2003, available from www.mikes.fi

NORDTEST (2004) Handbook for calculation of measurement uncertainty in environmental laboratories. Report TR 537 available from www.nordicinnovation.net/nordtest.cfm

13   Control Charts

Michael Koch and Michael Gluschke

Control Charts may be the most powerful tool to demonstrate and to assure quality in chemical measurements. Therefore they are widely used in all kinds of laboratories and it is hard to imagine quality management systems in laboratories without control charts.

Slide 1

ISO/IEC 17025 requires quality control procedures and, where practicable, the use of statistical techniques to monitor the validity of tests and detect possible trends.

Assuring the Quality of Test and Calibration Results - ISO/IEC 17025 – 5.9

- *The laboratory shall have quality control procedures for monitoring the validity of tests and calibrations undertaken.*
- *The resulting data shall be recorded in such a way that trends are detectable and, where practicable, statistical techniques shall be applied to the reviewing of the results.*

Slide 2

These actions have to be planned and reviewed. ISO/IEC 17025 gives examples of such procedures:
- the use of reference materials (see chapter 14)
- participation in proficiency testing schemes (see chapter 15)
- replicate tests
- retesting of retained items

Assuring the Quality of Test and Calibration Results - ISO/IEC 17025 – 5.9

- *This monitoring shall be planned and reviewed and may include, but not be limited to, the following:*
  - *regular use of certified reference materials and/or internal quality control using secondary reference materials;*
  - *participation in interlaboratory comparison or proficiency-testing programmes;*
  - *replicate tests or calibrations using the same or different methods;*
  - *retesting or recalibration of retained items;*
  - *correlation of results for different characteristics of an item.*

B.W. Wenclawiak et al. (eds.), *Quality Assurance in Analytical Chemistry: Training and Teaching*, DOI 10.1007/978-3-642-13609-2_13, © Springer-Verlag Berlin Heidelberg 2010

Slide 3

Control charts are a very powerful technique that can be used for the control of routine analyses. ISO/IEC 17025 requires the use of such statistical techniques wherever practicable.

Control Charts

- Powerful, easy-to-use technique for the control of routine analyses
- ISO/IEC 17025 demands use wherever practicable

Slide 4

Shewhart firstly introduced control charts in 1931 for the control of manu-facturing processes. Sudden occurring changes as well as gradual worsening of quality can easily be detected. Immediate interventions reduce the risk of producing rejects and minimize client complaints.

History

- Introduced by Shewhart in 1931
- Originally for industrial manufacturing processes
- For suddenly occurring changes and for slow but constant worsening of the quality
- Immediate interventions reduce the risk of production of rejects and complaints from the clients

Slide 5

The principle of control charts is very simple:
- Take samples during the process,
- measure a quality indicator and
- mark the result on a chart with limits.

Principle

- Take samples during the process
- Measure a quality indicator
- Mark the measurement in a chart with warning and action limits

Slide 6

If control charts are transferred to analytical chemistry the first thing to do is to assign the target value. If a reference material / certified reference material (RM/CRM) is used, the certified value can be used as the target value. This is advisable only, if the mean of the measurements is close to the reference value. Otherwise out-of-control situations would occur very frequently. So in most cases the arithmetic mean of the measurements is used as target value.

Control Charts in Analytical Chemistry

- Target value
  - Certified value of a RM/CRM
  - Mean of often repeated measurements of the control sample

Slide 7

For the warning and action limits there is a convention to use the ± 2s- and ± 3s-range respectively. If the data are normally distributed, 95.5% of the data are within $\mu \pm 2\sigma$ and 99.7% within $\mu \pm 3\sigma$. Since μ and σ are always unknown in analytical chemistry, the estimated values $\bar{x}$ and s are used.

Control Charts in Analytical Chemistry

- Warning / action limits
  - If data are normal distributed
  - 95.5% of the data are in $\mu \pm 2\sigma$
  - 99.7% are in $\mu \pm 3\sigma$
- $\bar{x} \pm 2s$ is taken as warning limits
- $\bar{x} \pm 3s$ is taken as action limit

Slide 8

The probability that 99.7 % of the "correct" measurement results are expected to lie within the action limits means that only 3 out of 1000 correct measurement results will lie outside this range. "Correct" means that they belong to the same statistical distribution of data. Therefore the probability is very high that a result that is outside the action range is incorrect.

If such a value is obtained the process should be stopped immediately and examined for errors.

Action Limits

- There is a probability of only 0.3 % that a (correct) measurement is outside the action limits (3 out of 1000 measurements)
- Therefore the process should be stopped immediately and searched for errors

Slide 9

The probability of correct data to be outside the warning limits is 4.5%. This is a reasonably high probability. Therefore such values should only give a warning signal and no immediate action is required.

Warning Limits

- 4.5% of the (correct) values are outside the warning limits.
- This is not <u>very</u> unlikely.
- Therefore this is only for warning, no immediate action required.

Slide 10

There are different standard deviations depending on the measurements conditions: repeatability conditions, between-batch and interlaboratory reproducibility conditions.
The measurements on the control sample usually are made with each batch and finally marked on the control chart are done under between-batch conditions. Therefore the standard deviation for the calculation of the limits should also be determined under the same between-batch conditions. The most common way to estimate this standard deviation is to use the results from a pre-period of about 20 working days. The use of the repeatability standard deviation would result in too narrow limits whilst interlaboratory conditions would lead to limits that are too wide.

Calculation of Standard Deviation

- Measurements marked in the control chart are between-batch
- Standard deviation should also be between-batch
- Estimation from a pre-period of about 20 working days
- Repeatability STD ➔ too narrow limits
- Interlaboratory STD ➔ too wide limits

Slide 11

Warning and action limits calculated in the above way should not be used without considering the "fit for the purpose" requirement. The limits may be too narrow or too wide for the analytical purpose. If this is the case the limits should be adjusted to such requirements.

Limits ⇔ Fitness for Purpose

- Action and warning limits have to be compatible with the fitness-for-purpose demands
- No blind use

Slide 12

The following slides will show some situations that are very unlikely to occur with normally distributed data and therefore should be treated as out-of-control situations.
This slide shows a value that is outside the action limit.

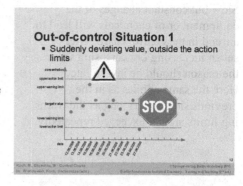

Slide 13

Although it is not very unlikely that a single value is outside the range defined by the warning limits, the probability that two out of three successive values are outside this range is very low. Hence this should be treated as an out-of-control situation as well.

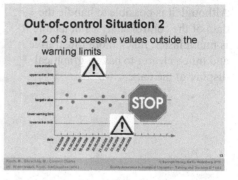

Slide 14

In analytical chemistry there are many possibilities that can lead to a system-atic shift in the measurement results. If seven successive values are on one side of the central line the process should also be stopped and the reason should be investigated. Please not that in this case the fitness-for-purpose criteria are still fulfilled since of all of the data are within the warning limits. So the situation is not as critical as in the previous situations. Nevertheless the data show that something happened to our dataset and we should investigate that.

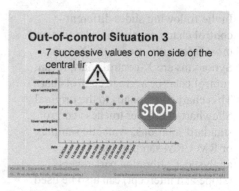

Slide 15

Slow but constant changes of the equipment or of chemicals will lead to a trend in the control chart. After seven increasing or decreasing values the reason should be investigated. Here the same applies as in the previous situation as long as the warning limits are not exceeded repeatedly.

Slide 16

Although it is possible to handle the data of the control analyses in a table it is much faster, much more illustrative and much clearer to have a graphical display of the data.

Slide 17

In the following slides different control charts are shown. The first and most frequently used is the X-chart. Synonyms are X-control chart, mean control chart or average control chart. This chart corresponds to the original Shewhart-chart. For trueness control, standard solutions, synthetic samples or RM/CRM samples may be analysed. Calibration parameters (slope and intercept) can also be used in a X-chart to check the constancy of the calibration.

Slide 18

The Blank Value Chart is also very important. This is a special form of a Shewhart chart where the direct measurements (e.g. in Volts) from the analyses of blank samples are used. From this chart information can be received about contamination of reagents e.g. from the environment and the state of the analytical system.

Different Control Charts
Blank Value Chart

- Analysis of a sample, which can be assumed to not contain the analyte
- Special form of the X-chart
- Information about
  - The reagents
  - The state of the analytical system
  - Contamination from environment
- Enter direct measurements, not calculated values

Slide 19

The Recovery Rate Chart reflects the influence of the sample matrix. Values from the analysis of a spiked and non-spiked sample are used.

Different Control Charts
Recovery Rate Chart - I

- Reflects influence of the sample matrix
- Principle:
  - Analyse actual sample
  - Spike this sample with a known amount of analyte
  - Analyse again
- Recovery rate:

$$RR = \left(\frac{x_{spiked} - x_{unspiked}}{\Delta x_{expected}} \right) \cdot 100\%$$

Slide 20

Proportional systematic errors are detected with a Recovery Rate Chart, but not constant systematic errors (e.g. too high blank values). Additionally the spiked analyte might be bound to the matrix differently. This possibly results in a higher recovery rate for the spike than for the originally bound analyte.

The mean value (usually used as the target value) should be around 100%.

Different Control Charts
Recovery Rate Chart - II

- Detects only proportional systematic errors
- Constant systematic errors remain undetected
- Spiked analyte might be bound differently to the sample matrix → better recovery rate for the spike
- Target value: around 100%

If the mean deviates the laboratory has to decide on the background f the fitness-for-purpose if this deviation is acceptable.

Slide 21

Often analyses are carried out more than once. In this case the range between the repeated measurements can also be used in a control chart. This is the Range Chart. It can be used only as a repeatability precision check, because the target value for the analysis is not known. This chart only has upper limits. It can be used with the range itself (R-chart) or with the percentage difference (R%-chart).

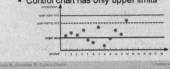

Different Control Charts
Range Chart

- Synonyms are R-chart or Precision chart.
- Absolute difference between the highest and lowest value of multiple analyses
- Precision check
- Control chart has only upper limits

The latter is used if the percentage range is expected to be more constant in the respective concentration range than the range itself.

Slide 22

The Difference Chart is very similar to the range chart but it uses the difference between double measurements together with its sign. The sample is measured at the beginning and at the end of a series. The difference between the two measurements is marked in the control chart with its sign.

Different Control Charts
Difference Chart - I

- Uses difference with its sign
  - Analyse actual sample at the beginning of a series
  - Analyse same sample at the end of the series
- Calculate difference (2nd value – 1st value)
- Mark in control chart with the sign

Slide 23

The target value of a Difference Chart should be around zero. Otherwise there is a drift in the analyses during the series and again the laboratory has to decide if this is acceptable. If not, the calibration of the method has to be made more frequently.
A Difference Chart gives information about precision and drift.

Different Control Charts
Difference Chart - II

- Target value: around 0
  - Otherwise: drift in the analyses during the series
- Appropriate for precision and drift check

Slide 24

The Cusum Control Chart is a very special chart from which a lot of information can be drawn. Cusum is the abbreviation for <u>cu</u>mulative <u>sum</u> and means the sum of all differences from the target value. Every day the difference of the control analysis from the target value is added to the sum of all the previous differences.

Different Control Charts
Cusum Chart - I

- Highly sophisticated control chart
- Cusum = cumulative sum = sum of all differences from one target value
- Target value is subtracted from every control analyses and difference added to the sum of all previous differences

Slide 25

This slide gives an example of the use of a Cusum Chart. The upper chart is a conventional X-chart, the lower one a Cusum Chart. Starting from the 11th value, all values are below the target value originating from slow between-batch-drift in the analyses. This can be seen in the Cusum Chart by a descending cumulative sum.

Different Control Charts - Cusum Chart - II

T = 80	s = 2.5		
Nr.	x	x-T	Cusum
1	82	+2	+2
2	79	-1	+1
3	80	0	+1
4	78	-2	-1
5	82	+2	+1
6	79	-1	0
7	80	0	0
8	79	-1	-1
9	78	-2	-3
10	80	0	-3
11	76	-4	-7
12	77	-3	-10
13	76	-4	-14
14	76	-4	-18
15	75	-5	-23

Slide 26

For the evaluation of a Cusum Chart a V-mask is used. This mask is laid on the chart so that the last marked value has a distance of d from the peak. With the correct choice of the distance d and the angle Θ, the relation between false positive and false negative alarms can be adjusted for an individual optimum.

Different Control Charts - Cusum Chart - III

- V-mask as indicator for out-of-control situation

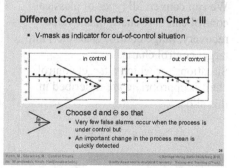

in control   out of control

- Choose d and Θ so that
  - Very few false alarms occur when the process is under control but
  - An important change in the process mean is quickly detected

Slide 27

The Cusum Chart shows very clearly the point from which the process ran out of control. The average run length, i.e. the time needed to detect an out-of-control situation is shorter than for other control charts. Furthermore, the size of a change in the process can be detected from the slope of the chart.

Different Control Charts
Cusum Chart - IV

- Advantages
  - It indicates at what point the process went out of control
  - The **average run length** is shorter
    - Number of points that have to be plotted before a change in the process mean is detected
  - The size of a change in the process mean can be estimated from the average slope

Slide 28

Target control charts are control charts with fixed quality criterions. In the contrary to classical control charts of the Shewhart-type these control charts operate without statistically evaluated values.
The limits for this type of control charts are given by external prescribed and independent quality criterions.

Different Control Charts
Target Control Charts - I

- In the contrary to classical control charts of the Shewhart-type the target control charts operates with fixed quality criterions and without statistically evaluated values
- The limits for this type of control charts are given by external prescribed and independent quality criterions (fitness for purpose)

Slide 29

We can convert all types of classical control chart (X-chart, blank value, recovery, range control chart etc.) into target control charts.
Situations for which a target control chart is appropriate are described in this slide.

Different Control Charts
Target Control Charts - II

- All types of classical control chart (X-chart, blank value, recovery, R-, R%-chart etc.) can be used as a target control chart
- A target control chart is appropriate if:
  - There is no normal distribution of the values from the control sample due to persisting out of control situations (e.g. blank values)
  - There are not enough data available for the statistical calculation of the limits (rarely analysed parameters)
  - There are external prescribed limits which have to be applied to ensure the quality of analytical values

Slide 30

The control samples for the target control charts and their measures are the same as for the classical control charts. The limits are given by different external quality criterion, described in the slide.

Different Control Charts
Target Control Charts - III

- The control samples for the target control charts are the same as for the classical control charts
- The limits might be given by
  - Requirements from legislation
  - Standards of analytical methods and requirements for internal quality control
  - The (minimum) laboratory-specific precision and trueness of the analytical value, which have to be ensured
  - The evaluation of laboratory-intern known data of the same sample type

Slide 31

The chart is constructed with an upper and lower limit. A pre-period is not necessary. The target control chart of the range and in some cases also of the blank only needs the upper limit. The analytical method is out-of-control, if the analytical value is higher or lower than the respective prescribed limits. Nevertheless trends in the analytical quality should be identified and steps should be taken against them.

Different Control Charts
Target Control Charts - IV

- Constructed with an upper and lower limit
- Pre-period is not necessary
- Out-of-control only, if the analytical value is higher or lower than the respective limit
- Nevertheless trends in the analytical quality should be identified and steps should be taken against them

Slide 32

This slide shows a target control chart for ammonia reference material measurements. The chart was constructed with the Excel® tool ExcelKontrol 2.1, which also is included in the electronic material to this book. In contrary to the classical control chart here only two limits and one out-of-control situation (outside control limits) are usually used.

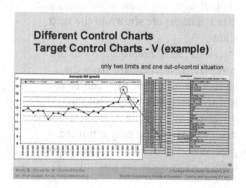

Different Control Charts
Target Control Charts - V (example)

only two limits and one out-of-control situation

Slide 33

This slide contains links to EXCEL-sheets from ExcelKontrol 2.1. This sheets can be used to run control charts with checks for out-of-control situations and with graphical displays. In the electronic material you will find the whole zipped program, which also allows target control charts.
Details and updates you will find in the internet (http://www.labcontrol-excel.org).

EXCEL-Tool for Control Charts ExcelKontrol 2.1

- X-/mean-charts
- Blank value chart
- Range chart with absolute ranges
- Range chart with relative ranges
- Recovery rate chart
- Differences chart

Slide 34

Control samples are a prerequisite for every control chart. They have to fulfil certain requirements that are shown in this slide.

Control Samples

- No control chart without control samples
- Requirements:
  - Must be suitable for monitoring over a longer time period
  - Should be representative for matrix and analyte conc.
  - Concentration should be in the region of analytically important values (limits!), if possible
  - Amount must be sufficient for a longer time period
  - Must be stable for several months
  - No losses due to the container
  - No changes due to taking subsamples

Slide 35

There are various possibilities for control samples. Their advantages and disadvantages are shown in the next slides.
Provided that standard solutions are completely independent from the calibration solutions they can be used to verify the calibration. But the influence of a sample matrix cannot be detected. Therefore only a limited precision check and only a very little control of trueness are possible.

Control Samples Standard Solutions

- To verify the calibration
- Control sample must be completely independent from calibration solutions
- Influence of sample matrix cannot be detected
- Limited control for precision
- Very limited control for trueness

Slide 36

Blank samples, i.e. samples that can be assumed to not contain the analyte, are necessary for blank value charts. With such samples it is possible to detect the errors listed in the slide.

Control Samples
Blank Samples

- Samples which probably do not contain the analyte
- To detect errors due to
  - Changes in reagents
  - New batches of reagents
  - Carryover errors
  - Drift of apparatus parameters
- Blank value at the start and at the end allow identification of some systematic trends

Slide 37

Real samples can be used for Range and Difference Charts. They deliver a rapid repeatability precision control, but no trueness check. If different matrices are analysed it might be useful to separate the results from different matrices in different charts.

Control Samples
Real Samples

- Multiple analyses for range and differences charts
- If necessary separate charts for different matrices
- Rapid precision control
- No trueness check

Slide 38

Real samples that are analysed with and without a spike are used for the Recovery Rate Chart. It is necessary to consider, whether the sample used for control has a representative matrix and whether the spike is bound to the matrix in the same way as the analytes in the sample.

Control Samples
Real Samples Spiked with Analyte

- For recovery rate control chart
- Detection of matrix influence
- If necessary separate charts for different matrices
- Substance for spiking must be representative for the analyte in the sample (binding form!)
- Limited check for trueness

Slide 39

Synthetically mixed samples are in very rare cases representative for real samples. Nevertheless, if they are, they can provide good checks on precision and trueness.

Control Samples
Synthetic Samples

- Synthetically mixed samples
- In very rare cases representative for real samples
- If this is possible ➜ precision and trueness check

Slide 40

CRMs would be the ideal control samples, but they normally are too expensive and very often not available. In-house reference materials that are regularly checked against a CRM under repeatability conditions are a good alternative.
If retained sample material from an interlaboratory test is available and it is stable after the test is finished, this also can be used as an in-house reference material.

Control Samples
Reference Materials

- CRM are ideal control samples, but
  - Often too expensive or
  - Not available
- In-house reference materials are a good alternative
  - Can be checked regularly against a CRM
  - If the value is well known ➜ good possibility for trueness check
- Sample material from interlaboratory tests

Slide 41

There are many possible control charts for each analysis. This leads to the questions, which is appropriate and how many control charts are necessary. There is no general answer to these questions except that the quality control measures have to be fit for the purpose of the analyses.
The laboratory manager has to decide, what he thinks is "fit for the purpose" (and the assessor in the accreditation process will assess this decision). But there are some general rules:

Which One?

- There are a lot of possibilities
- Which one is appropriate?
- How many are necessary?

- The laboratory manager has to decide!
- But there can be assistance

Slide 42

For frequently used methods control charts are absolutely necessary. If all the samples that are analysed have the same matrix, the sample preparation should be included in the control procedure. If the matrix changes very often, it could be useful to limit the procedure to the measurement step only.

Choice of Control Charts - I

- The more frequent a specific analysis is done the more sense a control chart makes
- If the analyses are always done with the same sample matrix, the sample preparation should be included. If the sample matrix varies, the control chart can be limited to the measurement only

Slide 43

Some standards or decrees include the obligatory measurement of control samples or repeated measurement. This can be used for control charts with only little effort. Other values like calibration parameters are also available without additional work. They also can be used for control charts, especially if the stability of calibration is known to be a weak point in the procedure.

Choice of Control Charts - II

- Some standards or decrees include obligatory measurement of control samples or multiple measurements. Then it is only a minimal additional effort to document these measurements in control charts
- In some cases the daily calibration gives values (slope and/or intercept) that can be integrated into a control chart with little effort

Slide 44

Control charts are a very powerful tool for quality control and should be used as a matter of course in every laboratory.

Benefits of Using Control Charts

- A very powerful tool for internal quality control
- Changes in the quality of analyses can be detected very rapidly
- Good possibility to demonstrate ones quality and proficiency to clients and auditors

Bibliography:

Duncan AJ (1965) Quality Control and Industrial Statistics, R.D. Irwin Inc., Homewood, Illinois

Farrant TJ (1997) Practical Statistics for the Analytical Scientist – a Bench Guide. Royal Society of Chemistry for the Laboratory of the Government Chemist (LGC), Teddington

Funk W, Dammann V, Donnevert, G (2006) Quality Assurance in Analytical Chemistry. 2nd edition, Verlag Wiley VCH Weinheim

ISO 8258:1991 – Shewhart control charts

Miller JC, Miller JN (2005) Statistics and Chemometrics for Analytical Chemistry. 5th edition, Pearson Education, Harlow

Montgomery DC(1985) Introduction to Statistical Quality Control, Wiley New York

NORDTEST: Internal Quality Control – Handbook for Chemical Laboratories. NORDTEST Report TR 569, available from www.nordicinnovation.net/nordtest.cfm.

Shewhart W (1931) The Economic Control of Quality of Manufactured Product. D. van Nostrand Company, New York

Wernimont G (1946) Use of Control Charts in the Analytical Laboratory. Industrial and Engineering Chemistry, Analytical Edition 18, 587-592

14 (Certified) Reference Materials

Ioannis Papadakis

This chapter gives an overview of the role of reference materials in quality assurance in analytical laboratories.

Slide 1

The presentation gives some definitions that are important to the selection use and production of reference materials and provides documentation for further reading and study. There is discussion on the different types of reference materials with some insight into their production process. This is followed by information relevant to

Index

- Definitions - documentation
- Types of (C)RMs
- Production of (C)RMs
- Selection and use of (C)RMs
- CRMs and traceability
- Suppliers/producers of (C)RMs

selection and use of reference materials with a discussion on the relation between reference materials and traceability, which is very important in the context of this presentation. Finally some internet addresses are given of the major certified reference materials producers.

Slide 2

Over the last years a lot of work has been done in the reference materials area by the International Organization of Standardization (ISO).
A special committee (ISO-REMCO) prepared many guidance documents relevant to reference materials.
A list of these 'guides' is given here.
This list is quite exhaustive starting

ISO Guides

- ISO 30 (1992) terms and definitions used in connection with RMs
- ISO 31 (2000) certificates of RMs
- ISO 32 (1997) calibration in analytical chemistry and use of CRMs
- ISO 33 (2000) uses of CRMs
- ISO 34 (2009) general requirements for the competence of RM producers
- ISO 35 (2006) certification of RMs

from 'terms and definitions' and going into details such as 'certification'.
These documents are very useful, covering most aspects of reference materials.

B.W. Wenclawiak et al. (eds.), *Quality Assurance in Analytical Chemistry: Training and Teaching*, DOI 10.1007/978-3-642-13609-2_14, © Springer-Verlag Berlin Heidelberg 2010

Slide 3

Let's see now the ISO definition of a reference material. The definition is concerned with sufficiently homogeneous and well-established property values. These values can be used for calibration of an apparatus, assessment of a measurement method or for assigning values to other materials.

Reference Material Definition

Material or substance one or more of whose property values are sufficiently homogeneous and well established to be used for the calibration of an apparatus, the assessment of a measurement method, or for assigning values to materials

Slide 4

There also is a note in the definition, which states the obvious and says that a reference material can be in all three possible forms, i.e. gas, liquid or solid. However the second part of the note is very interesting, because it gives some examples. Water can be considered as a reference material for the calibration of a viscometer since for a given temperature its viscosity is homogeneous and well established. In an analogous way sapphire can be used as heat-capacity calibrant in calorimetry and in analytical chemistry various solutions are used for calibration of analytical instruments.

Reference Material (RM)

- A reference material may be in the form of a pure or mixed gas, liquid or solid
- Examples are water for the calibration of viscometers, sapphire as a heat-capacity calibrant in calorimetry, and solutions used for calibration in chemical analysis

Slide 5

But what about the certified reference materials? This is a group of reference materials with a weighty name and there is a lot of discussion about them. The ISO definition is given here. In general they are reference materials, thus they also have property values sufficiently homogeneous and well established. However this is not enough. There should also be a certificate and the property values should be certified by a procedure that establishes traceability to an accurate realisation of the unit in which the property value is expressed. Moreover the certified property value should be accompanied by an uncertainty at a stated level of confidence.

Certified Reference Material Definition

Reference material, accompanied by a certificate, one or more of whose property values are certified by a procedure which establishes traceability to an accurate realisation of the unit in which the property values are expressed, and for which each certified value is accompanied by an uncertainty at a stated level of confidence

Slide 6

This definition is also accompanied by notes. The first says that certified reference materials are a sub-class of 'measurement standards' (or 'etalons'). The second gives us some information related to the usual production process of certified reference materials. Usually certified reference materials are prepared in batches and the property values (including the uncertainty) are determined in representative samples.

Certified Reference Material (CRM)

- All CRMs lie within the definition of "measurement standards" or "etalons"
- CRMs are generally prepared in batches for which the property values are determined within stated uncertainty limits by measurements on samples representative of the whole batch

To give an example, if a producer prepares 4000 vials from a batch of human serum, to be used as certified reference material for cholesterol measurements, it is neither practical nor realistic to measure the cholesterol content in all the 4000 vials. The measurements will be performed only on a number of vials. This will be discussed in more detail below.

Slide 7

Concerning the classification of reference materials, this usually is a matter of grouping. There is no hidden rule and it is usually done for presentation reasons. This slide presents several ways of classification, which are given in a book by Zschunke.

Classification of RMs

- Physical character
  - Gases, liquids or solids
- Supplied property
  - Pure chemical species, physico-chemical property
- Preparation method
  - Synthetic mixtures, natural materials
- Metrological qualification
  - Primary RM, secondary RM
- Intended use
  - Calibration of instruments, validation of analytical methods

Reference Materials in Analytical Chemistry - A Guide for Selection and Use, edited by A Zschunke, Springer, 2000

Slide 8

Passing from the theoretical part of the presentation to the more practical one, the production of certified reference materials will now be discussed. Certified reference materials are intentionally chosen because they are the most important for quality assurance. According to the ISO guide 35, the production of a certified reference material is an integrated process of correct preparation, demonstration of homogeneity and stability of the properties as well as accurate and traceable characterization of these properties. In addition all the components of uncertainty of "the sample on the desk of the user" should be accounted for.

CRM Production
According to ISO-35

- Producing a CRM is:
  - The integrated process of correct preparation, homogeneity and stability demonstration, and accurate and traceable characterisation,
  - Whereby all components of uncertainty of "the sample on the desk of the user" should be properly accounted for according to the ISO-GUM (Guide to the expression of uncertainty in measurement)

Slide 9

This does not sound to be very complicated, but it certainly is not a trivial job. It requires special skills and sophisticated installations in order to process materials in a suitable form, especially matrix certified reference materials. Imagine a few thousand bottles of freeze-dried orange juice with demonstrated homogeneity and stability for metals content! In addition there is a need for demonstrated measurement capability to produce the reference value, in other words to being able to adequately demonstrate the measurement traceability and measurement uncertainty.

CRM Production
Non-trivial Job!

- It requires ...
  - Skills and installation to process the material in a suitable form, especially for matrix CRMs
    (e.g. making 2000 containers of orange juice sample for heavy metals content measurement, with demonstrated homogeneity and stability)
  - Demonstrated measurement capability, to produce reference value

Slide 10

The list presented here divides the production process into different steps. This is not the only way of dividing the production process and it does not mean that all the steps need to be followed in every case. This is a detailed description according to the perception of the author.

CRM Production
Production Process

• Select the material
• Prepare the units (e.g. bottles)
• Labelling
• Measure homogeneity
• Measure stability
• Assignment of reference values
• Distribution/marketing

Slide 11

At the beginning of the production process a decision needs to be taken concerning the proposed certified reference material. There should be a specific need in the analytical commu- nity in terms of measurement, which needs to be met and of course there should be material available in a suit- able form, before taking the decision to start the process.

Select the Material

• Depends on the needs of the analytical community
• Measurement that needs to be supported
• Depends on availability of material

Slide 12

The starting material (batch) after appropriate homogenization should be stored in appropriate containers. Later these containers would be distributed to the interested scientists in order to support their measurements. Of course, first, the appropriate containers, in terms of size (large, small), shape (bottle, vial), properties (hard, soft,

Prepare the Units

• Select appropriate container (size, shape, material, properties ...)
• Prepare the units (e.g. vials) under the appropriate conditions
• Each unit should contain the appropriate amount of material
• Prepare appropriate number of units

coloured), material (glass, plastic) have to be selected. Some other important items at this stage are the preparation of the units under the appropriate conditions (e.g. freeze-dried material under low humidity), each unit should contain an appropriate amount of material (depending on the amount needed for each measurement and the availability of the material) and the appropriate number of units has to be decided (taking into account the needs for this specific certified reference material).

Slide 13

One of the important steps in the production process is the measurement of the homogeneity. This step ensures that all the units of the certified reference material carry the same (within the uncertainties) property value. Usually for this purpose a rapid measurement method is used, which requires only a small sample quantity and gives good reproducibility. The reasons for the above is that a lot of

Measure Homogeneity

- Fast measurement method
- Small sample quantity
- Good reproducibility
- Within containers
- Between containers

measurements need to be performed (rapid method), the total sample quantity is given (small quantity) and there is no need for very accurate measurements since the certification measurements are performed at a later stage (but reproducibility has to be good). It must be mentioned that two 'kinds' of homogeneity are usually measured, one within each container and one between different containers.

Slide 14

Another step of the production process, equally important to the homogeneity measurement, is the stability measurement. The measurement method requirements are the same as those for the measurement of homogeneity for the same reasons. At this stage there are two additional requirements. The measurements are performed on samples stored at different temperatures, in order to determine the appropriate

Measure Stability

- Fast measurement method
- Small sample quantity
- Good reproducibility
- Different storing temperatures
- Different times

storage temperature and at different times, in order to determine the lifetime of the certified reference material.

Slide 15

Probably the most important step in the production process of a certified reference material is the assignment of the reference value(s). This, in the case of a reference material, is done by designation but in the case of a certified reference material by measurement. Even when using synthetic preparation, measurements are still needed for confirmation of the preparation process.

Assignment of Reference Value

- By measurement (CRM)
- By designation (RM)

Slide 16

Focusing on the assignment of the reference value by measurement, there are several ways to achieve this. It can be done either by using one measurement method (usually this should be a primary method in order to be able to demonstrate measurement traceability and measurement uncertainty), or use of one method by several laboratories or several methods by one or several laboratories.

Assignment of Reference Value by Measurement

- One method in one lab (e.g. a primary method)
- One method in several labs
- Several methods in several labs

Slide 17

The traceability of the reference value needs to be stated, but more importantly also demonstrated. In addition all the relevant information should be available in the certification report of the certified reference material.

Traceability of the Assigned Value

- Needs to be stated
- ... and demonstrated
- Information available in certification report

Slide 18

The uncertainty of the assigned value, as stated before, is also very important. Uncertainty of the assigned value is the uncertainty of the average concentration of 1 unit (of the n prepared) after storage for time T and after transport, i.e. when the certified reference material arrives at the desk of the user. In general this can be obtained by combining the uncertainties of the value assignment measurement, the uncertainty of the bottle to bottle variation (homogeneity), the uncertainty of the long-term stability (lifetime) and the uncertainty of the short-term stability (transport).

Uncertainty of the Assigned Value (U_{CRM})

U_{CRM} uncertainty of the average concentration of 1 unit (of the *n* prepared) after storage for time T and after transport

$$U_{CRM} [\%] = k*(u^2_{assign} + u^2_{bb} + u^2_{lts} + u^2_{sts})^{1/2}$$

u_{assign} uncertainty value assignment measurement
u_{bb} uncertainty bottle to bottle variation
u_{lts} uncertainty long term stability
u_{sts} uncertainty short term stability (transport)

Slide 19

Continuing with the production steps, labelling of the certified reference material is important in order to avoid confusion. Very often this step is not taken seriously with unfortunate results. It is important to have clear and concise labelling.

Labelling

- Equally important
- Clear and concise

Slide 20

The final step of the production is the marketing of the certified reference material and the distribution. The most important item at this stage is the delivery of the certified reference material together with a detailed certification report with all the information relevant to the production of the material as well as advice for the intended use. The producer should also ensure that the certified reference material is

Distribution / Marketing

- Certification report
- Appropriate price
- Appropriate delivery
- Appropriate storage
- Reach scientific community

available at an appropriate price, that the delivery is done in an appropriate way and at an appropriate time. Also the certified reference material should be stored under appropriate conditions, at all times until it arrives at the customer. Finally it must be ensured that the broad scientific community is aware of its availability.

Slide 21

Summarizing the production and in order to help the selection of high quality certified reference materials, in general the traceability of the certified property should be stated, the uncertainty according to ISO guidelines should be stated, both of the above should also be demonstrated and it is preferred that the certified reference material is produced according to ISO guide 35.

In General High Quality CRMs Should...

- State traceability of certified value
  (e.g. traceability to SI, or to values obtained with method XYZ)
- State an ISO-GUM uncertainty of certified value
- Demonstrate traceability & uncertainty of certified value
  (e.g. in a certification report; experimental evidence of demonstrated capability from participation to international comparisons)
- Preferably be produced according to the guidelines of ISO-35

Slide 22

As far as the use of the certified reference material is concerned, in an ideal situation the supplier or producer should give advice on the appropriate use to the user. In addition it is essential to obtain information relevant to storage temperature, influence of moisture and influence of contamination. The possibility of dividing the certified reference material into different portions after opening in order to maintain its original properties should always be considered.

Use of CRMs

- (Ideally) supplier should give advice
- Storage temperature
- Influence of moisture on long term stability (e.g. biological activity)
- Influence of contamination
- Possibility to divide in different portions after opening

Slide 23

Which are the main ways that a certified reference material can contribute towards better measurements? The usual answers to this question are three: Calibration, Validation and Measurement Control. Let's investigate these three one by one.

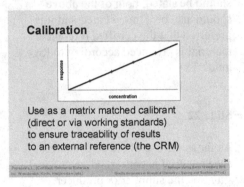

Use of CRMs

How can CRMs help my measurements?
- Calibration (?)
- Validation (?)
- Measurement control (?)

Slide 24

Starting with the calibration, the certified reference material can be used as a calibrant (directly or via working standards prepared using them) and in this way traceability of the results to the property values of the certified reference material is ensured. Thus this is a very 'valid' way for certified reference materials to improve the measurements in a laboratory.

Calibration

Use as a matrix matched calibrant (direct or via working standards) to ensure traceability of results to an external reference (the CRM)

Slide 25

In the case of validation, a certified reference material can be used to verify the performance or accuracy of a given measurement method. Questions such as "is there any method specific bias?" or "is there any systematic error?" can easily be answered in this way. Thus this is another useful way of using a certified reference material. In the case shown here certainly there is a significant bias.

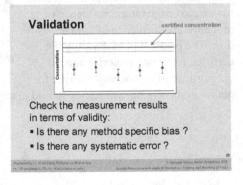

Validation

Check the measurement results in terms of validity:
- Is there any method specific bias ?
- Is there any systematic error ?

Slide 26

How about measurement control?
Usually the measurement control
process involves identical samples,
which are introduced into the sample
measurement sequence in order to
monitor the stability of the system.
Use of certified reference materials for
this purpose is possible, but actually it
is not an appropriate use of the material.
It is better and more economical to use
in-house materials of adequate homogeneity and stability (see chapter 13).

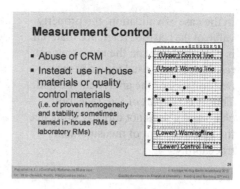

Slide 27

Thus summarizing the certified reference material should be used for calibration of an instrument or validation
of a measurement method but preferably
not used for measurement control.

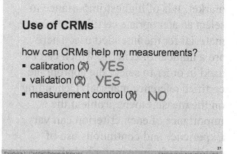

Slide 28

There is a lot of discussion about the
use of certified reference material in
terms of measurement traceability.
Let's see what is the influence for the
two cases where the certified reference
materials are useful in the laboratory
(i.e. calibration and validation). In the
case of calibration, the property value
of the certified reference material is
used to calibrate the analytical instru-

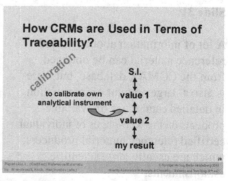

ment used for the measurement, thus it is used in order to obtain the measurement
result. In this way the property value of the certified reference material is part of
the traceability chain, as shown in the slide, and is directly involved in the establishment of the measurement traceability.

Slide 29

In the case of validation, the property value of the certified reference material is not used to obtain the measurement result. It is only used to verify the performance of the analytical method. Thus it is not part of the traceability chain and it does not directly influence the establishment of measurement traceability.

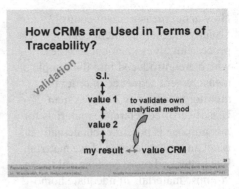

Slide 30

There is a large number of certified reference materials available in the market. It is of highest importance to select an appropriate certified reference material for the intended use. There are a number of criteria, which can be used in order to select the appropriate certified reference material. Depending on the measurement problem the importance of each criterion can vary. Experience and continuous use of certified reference material help to make a correct decision.

Selection of CRMs
- Availability (problem with matrix CRMs)
- Concentration range of certified property
- Uncertainty of certified property
- Traceability of certified property
- What is your uncertainty requirement
- Contribution of CRM uncertainty on your measurement uncertainty
- Demonstrated competence of CRM producer
- CRM matrix
- Cost

Slide 31

A lot of information about certified reference material can be obtained from the COMAR database, but there is also a large amount of information in detailed catalogues of available products on the web-pages of individual certified reference material producers, which are usually large scientific establishments.

CRM Producers

- General
  - COMAR database: http://www.comar.bam.de
- Individual suppliers
  - IRMM: http://www.irmm.jrc.be
  - BAM: http://www.bam.de
  - LGC: http://www.lgc.co.uk
  - NIST: http://www.nist.gov
  - others...

Slide 32

The general statement "I used a certified reference material, therefore my measurement result is correct" is very often used. This is an incorrect statement since when using a certified reference material there are still many things that can go wrong and which can result in an unreliable outcome.

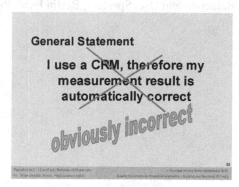

Slide 33

Summarizing this reference material chapter, you must remember that they are very important components for quality assurance in analytical laboratories. Their production is not an easy job, their selection and use requires specific information and careful decisions have to be made and finally the use of certified reference materials cannot replace careful laboratory work.

Bibliography

ISO Guide 30:1992 Terms and definitions used in connection with reference materials

ISO Guide 31:2000 Reference materials – Contents of certificates and labels

ISO Guide 32:1997 Calibration in analytical chemistry and use of certified reference materials

ISO Guide 33:2000 Uses of certified reference materials

ISO Guide 34:2009 General requirements for the competence of reference material producers

ISO Guide 35:2006 Reference materials – General and statistical principles for certification

Zschunke A (2000) Reference Materials in Analytical Chemistry - A Guide for Selection and Use, Springer, Berlin Heidelberg

15   Interlaboratory Tests

Michael Koch

Interlaboratory tests are one of the most important external quality assurance tools to assure comparability with all the other laboratories.

Slide 1

For interlaboratory tests sub-samples are distributed to different laboratories for concurrent testing.

What are Interlaboratory Tests?

- Randomly selected sub-samples from a source of material are distributed simultaneously to participating laboratories for concurrent testing

Slide 2

There are different reasons for inter-laboratory tests. One is method valida-tion, e.g. prior to the standardization of characterization of reference materials, which have to be certified. The third and most important for quality assurance is proficiency testing of laboratories. There are different requirements for each type of interlaboratory test.

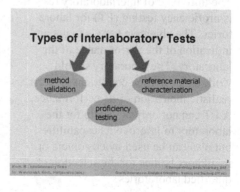

Types of Interlaboratory Tests

method validation

proficiency testing

reference material characterization

B.W. Wenclawiak et al. (eds.), *Quality Assurance in Analytical Chemistry: Training and Teaching*, DOI 10.1007/978-3-642-13609-2_15, © Springer-Verlag Berlin Heidelberg 2010

Slide 3

The objective for interlaboratory tests for method validation is the best possible characterization of the method. Therefore it has to be assured that every participating laboratory uses exactly the same method.

Interlaboratory Tests for the Validation of a Method

- Objective: best possible characterization of the method
- Laboratories have to use exactly the same method
- Assistance should be given to assure this

Slide 4

Interlaboratory tests for characterization of a reference material are different. It is very important that only very experienced laboratories take part in this test to get the best possible estimation of the "true" value.

Interlaboratory Tests for Characterization of a Reference Material

- Objective: best possible estimation of the "true" value of the concentration
- Concentration of the analyte in the material must be analysed by **experienced** laboratories
- Less experienced laboratories should not be allowed to participate

Slide 5

The third type of interlaboratory tests is proficiency testing (PT) for laboratories. The objective here is to get an indication of the performance of the laboratory. The laboratory should work under routine conditions to get a realistic indication of its performance. A PT can not only be a help for the laboratory to improve its capabilities, but also can be used by customers or regulatory bodies for the selection of qualified laboratories.

Interlaboratory Tests for Proficiency Testing of Laboratories

- Objective: to get an indication of the performance of an individual laboratory or a group of laboratories as a whole
- Laboratories should work under routine conditions
- Help for the laboratory to improve its quality
- Can be used by customers or regulatory bodies for the selection of qualified laboratories

Slide 6

Interlaboratory tests are a test for accuracy. Inaccuracy arises from systematic and random effects that are related to bias and precision respectively. As a result of a PT the laboratory should be able to determine, whether imprecision or bias is the reason for its inaccuracy.

Objectives of Proficiency Tests

- Basic concern is accuracy
- Inaccuracy contains systematic and random effects
- Laboratory can determine, whether imprecision or bias is the reason for its inaccuracy

Slide 7

Why should a laboratory participate in PT schemes? First there is the possibility to uncover errors that couldn't be found with other quality control measures and second a successful participation can be used as a certificate of competence for clients, authorities and accreditation bodies.

Motivation for the Laboratories

- To uncover errors that couldn't be found with internal quality control
- Use as certificate for competence in this testing field for clients, authorities and accreditation bodies

Slide 8

However there are also limitations. First of all it takes time to evaluate a PT. So it is quite a long time before the laboratory knows the result. The interlaboratory test is retrospective. It is therefore not possible to base quality management only on proficiency tests. Normally PTs cover only a small part of the laboratory's business. This is the second limitation. A PT analysis normally does not reflect the routine analysis, because the effort given to the former can be much greater than the effort given to the latter.

Limitations

- Interlaboratory tests are always retrospective
  - Organisation, distribution of samples, analyses, evaluation take time
  - It is dangerous to rely only on interlaboratory tests
- Proficiency tests cover only a small fraction of the often wide variety of analyses
- Proficiency tests do not reflect routine analyses

Slide 9

There are some guidelines on the requirements for proficiency testing schemes. The most important is the new ISO/IEC 17043 "Conformity assessment – General requirements for proficiency testing". It describes the development and operation of a PT scheme with all quality management requirements related to that. The first version of the "International Harmonized Protocol for the Proficiency Testing of (Chemical) Analytical Laboratories was a result of a joint workshop of IUPAC, ISO and AOAC and was published in 1993. A 2nd , revised version was published in 2006. This protocol also sets out the requirements for the provider and for the PT scheme itself.

Standards and Guidelines for Proficiency Testing - I

- ISO/IEC 17043:2010 – Conformity assessment – General requirements for proficiency testing
- IUPAC, ISO, AOAC (2006): The International Harmonized Protocol for the Proficiency Testing of Analytical Chemistry Laboratories

Slide 10

ILAC describes requirements for the competence of PT providers in its guideline G13. Due to the recent publication of ISO/IEC 17043, a withdrawal of ILAC G-13 is to be expected.
In ISO 13528 details on possible statistical methods for the evaluation of proficiency tests are given.
The following slides are mainly based on the content of these four guides and standards.

Standards and Guidelines for Proficiency Testing - II

- International laboratory accreditation cooperation (ILAC-G13:2007): Guidelines for the requirements for the competence of providers of proficiency testing schemes.
- ISO 13528:2005 - Statistical methods for the use in proficiency testing by interlaboratory comparisons.

Slide 11

Since interlaboratory tests are a great logistical challenge, especially if there are a lot of participants, the personnel of the provider must have special organizational capabilities. To under-stand the problems associated with the analyses there must be technical

Demands on the Provider - Personnel

- Special organisational capabilities
- Technical experts for the analysis
- Statisticians
- All staff have to be competent for the work it is responsible for

experts as well as statisticians involved in the evaluation. As in a good laboratory everybody has to be competent in the work for which he is responsible.

Slide 12

It should be a matter of course carefully to prepare an interlaboratory test. This includes the documentation of the plan before the start of the test.

Demands on the Provider – Planning

- The interlaboratory test should be carefully prepared.
- The planning must be documented before the start of the test

Slide 13

The minimum content of this plan is listed in the next three slides.

Demands on the Provider – PT Plan - I

- Name and address of the PT provider
- Name, address and affiliation of the coordinator and other personnel
- Activities to be subcontracted and the names and addresses of subcontractors
- Criteria to be met for participation
- Number and type of expected participants
- Selection of the measurand(s) or characteristic(s) of interest, including information on what the participants are to identify, measure, or test for
- Description of the range of values or characteristics, or both, to be expected for the PT items
- Potential major sources of errors involved in the area of PT offered

Slide 14

Demands on the Provider – PT Plan - II

- Requirements for the production, quality control, storage and distribution of PT items
- Reasonable precautions to prevent collusion between participants or falsification of results
- Description of the information which is to be supplied to participants and the time schedule for the various phases of the PT scheme
- Information on methods or procedures which participants need to use
- Methods to be used for the homogeneity and stability testing

Slide 15

> **Demands on the Provider –**
> **PT Plan - III**
>
> - Standardized reporting formats to be used by participants
> - Detailed description of the statistical analysis
> - Origin, metrological traceability and measurement uncertainty of any assigned values
> - Criteria for the evaluation of performance
> - Description of the data, interim reports or information to be returned to participants
> - Description of the extent to which participant results are to be made public
> - Actions to be taken in the case of lost or damaged proficiency test items

Slide 16

The data-processing equipment used by the provider also has to be fit for the purpose. This includes hardware as well as adequate software, which have to be verified and backed up.

> **Demands on the Provider –**
> **Data-processing Equipment**
>
> - Equipment should be adequate for
>   - Data processing
>   - Statistical analysis
>   - To provide timely and valid results
> - Software must be
>   - Verified and
>   - Backed up

Slide 17

The selection and preparation of the samples very often is the most crucial item in an interlaboratory test. All characteristics that could affect the integrity of the test have to be considered. This includes homogeneity and stability, which are dealt with in detail later. The PT provider has to look for possible changes in and the effect of ambient conditions on the samples.

> **Demands on the Provider – Test Item Preparation and Management - I**
>
> - For the selection of the test item all characteristics that could affect the integrity of the interlaboratory comparison should be considered
>   - Homogeneity
>   - Stability
>   - Possible changes during transport
>   - Effects of ambient conditions (e.g. temperature)

Slide 18

The samples that are used for
proficiency tests should be similar to
routine samples in order to get an
information about the routine capabilities of the laboratory. In some cases it
is useful to distribute a larger amount
of sample than is needed for the
determination, to give the laboratories
the opportunity to use this material as
an in-house reference material. But the
surplus can be also used to carry out
more analyses compared to routine, which is not desired.

**Demands on the Provider – Test Item
Preparation and Management - II**

- Samples used in the proficiency test
  should be similar to the samples that
  are routinely analysed in the
  laboratories
- Sample amount
  - Surplus of sample can be used as
    reference material
  - Surplus can be used to make excessive
    effort on the analyses

Slide 19

The PT provider has to take care that
all the distributed samples are comparable with each other. Therefore the
provider has to ensure the homogeneity
of the testing material before dividing
it into sub samples. The provider must
have a documented procedure for
homogenisation and for homogeneity
tests. It has to be assured that the
evaluation of the laboratory results is
not affected by inhomogeneities.

**Demands on the Provider -
Homogeneity - I**

- The PT provider has to ensure that every
  laboratory will receive samples that do not
  differ significantly in the parameters to be
  measured
  - Documented procedure for establishing this
    homogeneity
  - Degree of homogeneity ➔ evaluation of the
    laboratories results must not be significantly
    affected
- Any variation between the portions must be
  negligible in relation to the expected
  variations between the participants

Slide 20

This is easier for true solutions,
because they are homogeneous at a
molecular level. For solid samples this
is more difficult and special methods
must be applied to ensure homogeneity.
A procedure using ANOVA techniques
for the check for homogeneity is
described in the International
Harmonized Protocol.

**Demands on the Provider -
Homogeneity - II**

- True solutions are homogeneous at a
  molecular level
- For solid samples ➔ special care on the
  homogenisation
- A homogeneity check on the basis of
  ANOVA is described in the
  „International harmonized protocol..."

Slide 21

Another requirement for the PT samples is stability under the PT conditions for the time period between the sample production and the analyses in the laboratory, which has to be assured and tested by the PT provider.

Demands on the Provider - Stability - I

- Test material must be sufficiently stable
  - Under the conditions of storage and distribution to the participants
  - For the time period from producing the samples until the analyses in the participant's laboratory
- This stability has to be tested by the PT provider

Slide 22

The only way to test it is by making a repeat measurement after the estimated time necessary for the sample distribution has expired. There are two possible explanations for differences in the result:
- instability and
- between-batch variability of the provider's analyses.

Further information on stability can be derived from the provider's experience or from literature. It might also be necessary to perform accelerated tests under worst-case conditions.

Demands on the Provider - Stability - II

- Analysing a part of the samples after the estimated time necessary for the distribution
  - Differences in the results may be due to instability or to between-batch variability in the organiser's laboratory
- Information may be derived from the organiser's prior experience or obtained from technical literature
- Accelerated stability testing by worsening the ambient conditions for the samples

Slide 23

The PT provider has to assure that the changes in the samples are negligible and do not affect the evaluation of the laboratory performance.

Demands on the Provider - Stability - III

- The organiser has to ensure that the changes due to instability do not significantly affect the evaluation of the laboratories' performance

Slide 24

Normally the laboratory should use its routine method for the analysis of the PT samples, but there might be legal regulations that limit the choice of methods.
In order to conduct a method-specific evaluation and to be able to comment on the results of different methods, the PT provider should ask for the details of the method used.

Choice of Analytical Method

- Normally the laboratory should use its routine method
- The choice might be limited by e.g. legal regulations
- Organiser should ask for details
  - To conduct a method specific evaluation
  - To give comments on the methods used

Slide 25

A method-specific evaluation might look like that shown in this slide. The results obtained with the different methods are divided into five classes. Results which are between the range "assigned value" (av) $\pm$ 1 s, are in the class titled "correct", the range av $+$ 1s $< x \le$ av $+$ 2s is titled "high" and x $>$ av $+$ 2s "too high". A similar classification is made for "low values". In this figure differences between different

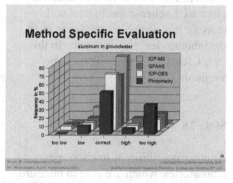

methods can easily be seen. But methods that show deviations like the photometry in the example are not necessarily bad methods. The example shows that there are more sources for possible mistakes. The consequence is that these mistakes are made by a number of analysts. But it definitely is possible to measure correct results with this method.

Slide 26

The performance of the laboratories is measured against the assigned value, which therefore is very important. This value has to be determined very carefully. If the assigned value is inappropriate the whole PT scheme is questionable. In principle it is the same

Determination of the Assigned Value

- One of the most critical features of a proficiency test
- Inappropriate value will drastically reduce the value of the scheme
- The same problem as in the certification of a reference material
  - But the organiser of a proficiency test cannot expend the same amount of effort

problem as the determination of the value of a reference material. However if the PT provider were to spend the same effort in the determination of the assigned value, the PT scheme would be much too expensive. In some cases it would not be possible to carry out such a detailed determination of the assigned value, because the samples are not stable enough for such a long procedure.

In the following slides various possibilities for the definition of the assigned value are shown.

Slide 27

A CRM is an ideal test material for a PT scheme. Unfortunately their high cost and lack of availability of suitable CRMs in the quantity and concentration range needed limit their use. Very often a PT scheme has to deal with more or less unstable samples to meet the laboratories' requirements. In this case CRMs are very unlikely to be available.

**Assigned Value –
Certified Reference Material**

- Ideal test material for a proficiency test
- Disadvantages
  - High costs
  - Limited availability
    - In the necessary quantity
    - And concentration range
  - CRM's have to be stable for months and PT often deals with more or less instable samples (foodstuffs, biomedical, environmental samples)

Slide 28

Another possibility is to choose expert laboratories, which are able to measure with high precision reference methods, utilising traceable calibration materials. If these laboratories use methods based on different physico-chemical principles and come to more or less the same result, it is very probable that the value is close to the "true" value.

**Assigned Value – Consensus of
"Expert Laboratories" - I**

- Mean of analysis by expert laboratories
  - With high precision reference methods and traceable materials for calibration
  - If different physico-chemical methods are used and the same results are obtained, it is more probable that the value is near to the „true" value

Slide 29

There are two disadvantages with this method. On the one hand much effort is required to ensure the accuracy of the reference measurements. This is a matter of capacity in the reference laboratories and a matter of costs as well. On the other hand there might be doubt among the participants about the assigned value, because "nobody is perfect". This is especially true if the mean of the participants' results deviates from the assigned value.

Assigned Value – Consensus of "Expert Laboratories" - II

- Disadvantages
  - Very much effort to ensure the accuracy of the reference measurements
  - "Nobody is perfect"
    - There might be doubts among the participants if the result of the expert laboratories deviates from the mean of the participants

Slide 30

Another method is spiking of materials with a known amount of the analyte. This can be done with high precision by gravimetric or volumetric methods. If there is no analyte in the base material the assigned value can be calculated directly from the added amount.

Assigned Value – Formulated or "Synthetic" Test Materials - I

- Materials, spiked with the analyte to a known extent
  - Can be made with extremely accurate amounts by gravimetric or volumetric methods
- If material does not contain significant amounts of the analyte
  - Assigned value directly from added amount
- If material contains analyte, this amount has to be characterized very well.

If the material already contains a certain amount of analyte, this amount has to be quantified with sufficient accuracy.

Slide 31

The disadvantages of this method are the difficulty of achieving sufficient homogeneity. Especially if the samples are solid materials (e.g. soil or foodstuff) this could be a significant problem. The second point is that the analyte originally contained in the material might be more tightly bound to the matrix than the added analytes.

Assigned Value – Formulated or "Synthetic" Test Materials - II

- Disadvantages
  - Difficult to achieve sufficient homogeneity, especially with solid materials
  - Analyte might be bound in a different chemical form
    - Especially in solid materials the originally contained analyte might be bound more strongly to the matrix

Then the recovery of the spiked analyte could be much better than that of the originally bound analyte.

Slide 32

The easiest and cheapest, and therefore most often used method is to take a consensus value of the participants as the assigned value. If the analysis being tested is very easy and straight-forward to perform, then this is not a problem since the assigned value will be a good estimate for the "true" value. For "conventional methods" this is the only possibility because the "true" value in this case is defined by the analytical procedure.

Assigned Value –
Consensus of Participants - I

- Easiest and cheapest way ➔ used very often
- If method for analysis is easy and straightforward ➔ good estimate of „true" value
- If a „convention method" (an empirically defined method) is used, the consensus value is the only possibility

Slide 33

But this method also has disadvantages. The consensus value could be seriously biased, if the majority of the participants have biased values in the same direc-tion, which could be the case e.g. in the analyses of highly volatile substances. Or there might be no real consensus at all. This could perhaps result from the use of two different methods, where one is biased.
All of these are common in trace analysis and therefore a participants' consensus is not always a good way for defining the assigned value.

Assigned Value –
Consensus of Participants - II

- Disadvantages
  - Consensus value might be seriously biased (e.g. analyses of highly volatile substances)
  - There might be no consensus at all
    - E.g. if two analytical methods are used, where one is biased
  - These circumstances are not uncommon in trace analysis
  - Care should be taken to decide whether a consensus value really is good choice

Slide 34

If a consensus value is used as the assigned value there are different possibilities to calculate it. If the arithmetic mean is used, an outlier test is required. But in many cases these tests are not very satisfactory, especially if several outliers are present. If the tests are strictly used, they can only be applied to normally distributed data, which is usually not the case in trace analysis.

Methods to Calculate Consensus
Value – Arithmetic Mean

- Requires an outlier test
  - But these tests are often not very satisfactory, especially if many outliers are present
  - Outlier tests assume normal distribution which is normally not true in trace analysis

Slide 35

The median is not affected by outlying data, but the statistical efficiency is not good. This means that the reliability of the estimation of a dataset's population mean from a small sample of data is lower than for the other methods.

Methods to Calculate Consensus Value – Median

- Not affected by outlying data
- Low statistical efficiency

Slide 36

"Robust Statistics" use trimmed data for the calculation of the estimated values. That means, that a part of the data set in the tails is excluded or modified prior to or during the calcula-tion. An easy example is the use of the interquartile range (the range between the first and the third quartile) instead of the whole data set.

Special methods have been developed in the past e.g. by Huber (algorithm A) and Hampel.

Methods to Calculate Consensus Value – Robust Mean

- "Trimmed" data; a certain part of the data on both tails of the data set is excluded prior to the calculation of the mean
- E.g. mean of interquartile range
  - Mean of data between the first and the third quartile of the data set
- Or algorithm A

Slide 37

The calculation of a robust mean according to a method introduced by Huber and described in the relevant standards is shown in this slide. The method starts with median as the initial value m for an iterative procedure. All data outside the m ± 1.5*STD are set to this border. Then a new value for m is calculated from the arithmetic mean of this new data. The procedure of transforming the initial data outside the range m ± 1.5*STD and calculation of a new mean is then repeated until there are no more changes in the values.

Methods to Calculate Consensus Value – Robust mean – Algorithm A

- Iterative process
  - Described in ISO 5725, ISO 13528 and the International Harmonized Protocol
  - Define initial value for m as median of all data
  - All data outside m ± 1.5 · STD are set to m + 1.5 · STD or m - 1.5 · STD
  - New value for m is calculated as arithmetic mean of the new data
  - Repeat until there are no changes

Slide 38

In the last years the robust statistical methods got more and more important. In all new relevant standards the use of these methods are now highly recommended.

Choice of Method to Calculate Consensus Value

- The use of robust methods (e.g. algorithm A) is highly recommended in all relevant standards

Slide 39

Once the assigned value is set the target value for the analyses is known. But it is also necessary to define a tolerance range for the analytical results. Values within this range are considered as satisfactory. This range might originate from a "fitness for purpose" requirement or it might be calculated from the dispersion of the data, i.e. from the standard deviation.

Performance Scoring

- Assigned value is the target
- For the assessment of laboratories a accepted range is necessary
  - Prescribed range originating from the demands put on the analysis (fitness for purpose)
  - Calculated from the standard deviation of the data set

Slide 40

Provided that the data are normally distributed, 95.5% of the values will lay within the range $\mu \pm 2\sigma$, and 99.7% within $\mu \pm 3\sigma$. That means in other words that on a confidence level of 95.5% all accurate data are inside $\mu \pm 2\sigma$.

Performance Scoring – Tolerance Range from STD

- Normally distributed set of data
  - 95,5% of the values inside a range of $\pm 2\sigma$
  - 99,7% of the values inside a range of $\pm 3\sigma$
- On a confidence level of 95,5 % all accurate data are inside $\mu \pm 2\sigma$

Slide 41

For proficiency tests z-scores have been widely used for many years. Z-scores represent the deviation from the assigned value in standard deviation units. The standard deviation may be calculated after exclusion of outliers or with robust statistics. In some cases it is set to a certain value according to the quality targets of the PT provider.

Performance Scoring – Z-score

- The deviation from the assigned value in standard deviation units

$$z - score = \frac{(x - \mu)}{s}$$

- The standard deviation is calculated after exclusion of outlier or with robust statistics

Slide 42

In the International Harmonized Protocol z-scores are classified into three categories. The range –2.0 to +2.0, corresponding to a confidence level of 95.5% is classified as satisfactory. The range 2.0 < |z-score| < 3.0 is classified as questionable since the probability that these data are accurate is only 4.5%. Data with |z-score| ≥ 3.0 are classified as unsatisfactory, because the confidence level is only 0.3%.

Performance Scoring – Classification in the Internat. Harmonized Protocol

- |z-score| ≤ 2.0 - satisfactory
- 2.0 < |z-score| < 3.0 - questionable
- |z-score| ≥ 3.0 - unsatisfactory

- Z-scores are common practise in the assessment of laboratory results

Slide 43

If the z-scores of the participants for a specific sample are sorted and depicted in a bar chart they normally show a typical S-shape. In this diagram the existence of a real consensus value can easily be seen. This figure also enables the participants to compare their results with their peers.

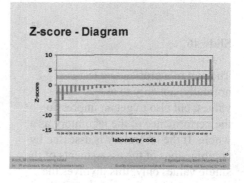

Z-score - Diagram

Slide 44

The concept of z-scores is based on the assumption that the data are normally distributed. But this is not true for data near the detection limit. The distribution has to be skewed; otherwise there would be a finite probability of negative values. For this reason tolerance limits should also be asymmetrical. If they are narrower below the assigned value, this also ensures that the lower tolerance limit will not be negative with the consequence that all data delivered with "< X" would have to be classified as acceptable.

Statistical distribution

- Data near to the limit of determination are not normal distributed
- Otherwise there should be negative values with a finite probability
- Tolerance limits should be asymmetrical (more narrow below the assigned value, more wide above it)

Slide 45

Two approaches attempt to solve this problem. The one is a transformation of the data to logarithms prior to the statistical calculations corresponding to a logarithmic normal distribution. The other is a modification of the z-scores with correction factors. This method was introduced first in a German standard for proficiency testing (DIN 38402 – 45), which in the meantime partially was transferred into ISO/TS 20612.

Solution Approaches for Asymmetrical Tolerance Limits

- Logarithmic normal distribution
  - Take the logarithm of the values prior to statistical calculations
- Modification of Z-scores

$$Z_a = \begin{cases} \frac{g}{k_1} \cdot Z \; if \; Z<0 \\ \frac{g}{k_2} \cdot Z \; if \; Z>0 \end{cases}$$ with g = quality limit for Z and k_1, k_2 = correction factors

$$\left(k_2+\frac{1}{v}\right)^2 \exp\left\{-\frac{1}{2}k_2^2\right\} = \left(-k_1+\frac{1}{v}\right)^2 \exp\left\{-\frac{1}{2}k_1^2\right\}$$

$$\left(1-\Phi\left(-\frac{1}{v}\right)\right)^2 \left(\Phi(k_2)-\Phi(-k_1)\right)=1-\alpha$$

v = rel. standard deviation
Φ = distribution function of standard normal distribution
1-α = confidence level (here: 0,955)
(from ISO/TS 20612:2007)

Slide 46

The aim of a proficiency test is normally, not only to assess single values, but also to get an impression of the performance of the laboratory as a whole. If the laboratory is assessed by combinations of the assessments of the single values only, this involves the danger of misinterpretation. This is

Laboratory Assessment

- By combination of single value assessment
- Involves danger of misinterpretation
  - A laboratory can measure one parameter permanently wrong, but nevertheless is positively assessed

because a laboratory can measure some of the parameters incorrectly all the time yet be positively assessed.

Slide 47

In the International Harmonized Protocol two possibilities for combined assessments are specified. For the rescaled sum of z-scores a combination of the z-scores is calculated from $RSZ=\Sigma z/\sqrt{m}$ with m=number of z-scores. This RSZ has the same scale as a z-score. But if all values are biased to the same direction, still being within the tolerance limits, this will lead to a negative total assessment. On the other hand if the values are biased in different directions, this improves the total assessment.

> **Combined Assessment According to Intern. Harmon. Protocol... - RSZ**
>
> - RSZ (rescaled sum of z-scores)
>   - RSZ = Σz/√m with m = number of scores
>   - Same scale as z-score
>   - Negative assessment, if all values are within the tolerance but a little biased in the same direction
>   - Errors with opposite sign cancel each other out

Slide 48

The sum of z-scores $SSZ=\Sigma z^2$ has a different scale, but doesn't consider the sign of the z-scores.

> **Combined Assessment According to Intern. Harmon. Protocol... - SSZ**
>
> - SSZ (sum of squared z-scores)
>   - Different scale, because X²-distributed
>   - Does not consider the sign of z-scores

Slide 49

Nevertheless in most PT schemes a combination of single value assessment is used, just counting positive and negative assessments of all values. This does not consider the value of the z-score.
In the PT scheme of the German water authorities the following limits are

> **Combination of Single Values Assessments**
>
> - Just counting positive and negative assessments of all values
> - The absolute value of the z-score is not considered
> - E.g. assessment in the proficiency tests of German water authorities
>   - 80 % of the values – |Z$_u$-score|≤2
>   - 80 % of the parameters successful

used for a positive total assessment: The tolerance limit for single values are set to |zU-scores| ≤ 2, 80% of the values have to be inside the tolerance limits, and 80% of the parameters have to be analysed successfully, meaning more than 50% of the values for this parameter are positively assessed.

Slide 50

When the proficiency test is finished the PT provider has to report the results to the participants. This should be done as soon as possible, i.e. normally not later than 1 month after the deadline for the return of the analytical results, to give a quick feedback to the participants. This enables the laboratories to react and take corrective actions. To maintain confidentiality, only a code should be used to identify the laboratory in the report.

Test Scheme Reports

- Should be distributed to the laboratories as soon as possible
  - Normally not later than 1 month after deadline for the analytical results
  - Laboratories need quick feedback for corrective actions
- Laboratories should be identified in the report by test specific codes to maintain confidentiality

Slide 51

The necessary content of a test scheme report is listed in ISO/IEC 17043 in detail and shown here in the next three slides.

Test Scheme Reports – ISO/IEC 17043 – Contents - I

- Name and contact details for the PT provider
- Name and contact details for the coordinator
- Name, function and signature of person authorizing the report
- Indication of which activities are subcontracted by the PT provider
- Date of issue and status (e.g. preliminary, interim, or final) of the report
- Page numbers and a clear indication of the end of the report
- Statement of the extent to which results are confidential
- Report number and clear identification of the PT scheme

Slide 52

Test Scheme Reports – ISO/IEC 17043 – Contents - II

- Clear description of the PT items used, including necessary details of the PT
- Item's preparation and homogeneity and stability assessment
- Participants' results
- Statistical data and summaries, including assigned values and range of acceptable results and graphical displays
- Procedures used to establish any assigned value;
- Details of the metrological traceability and measurement uncertainty of any assigned value
- Procedures used to establish the standard deviation for proficiency assessment, or other criteria for evaluation

Slide 53

**Test Scheme Reports –
ISO/IEC 17043 – Contents - III**

- Assigned values and summary statistics for test methods/procedures used by each group of participants (if different methods are used by different groups of participants)
- Comments on participants' performance by the proficiency testing provider and technical advisers
- Information about the design and implementation of the PT scheme
- Procedures used to statistically analyse the data
- Advice on the interpretation of the statistical analysis
- Comments or recommendations, based on the outcomes of the PT round

Slide 54

The participants in most schemes are interested in getting a certificate of successful participation in order to use it for advertising and demonstration of competence to their customers.

Certificate

- If the proficiency test scheme has regulations for the assessment of the laboratories on the basis of the data (successful / not successful) a certificate should be sent to the laboratory in case of successful participation.
- In many cases these certificates are used by the laboratories for demonstrating competence to their customers, i.e. for advertising.

Slide 55

This slide shows an example from a German PT scheme.

**Certificate –
Example**

Slide 56

For most laboratories it is essential
that their identity in a proficiency test
is kept confidential, because public
reports about poor performance could
ruin a laboratory. In the PT provider
organization the identity of the laboratories should also be known only to a
small number of persons, who must be
regularly instructed about their duty to
keep this information confidential. The
provider may be required to report the

Confidentiality
- Normally in all PT schemes the identity of all laboratories are kept confidential
- Public reports about poor performance of a laboratory in a proficiency test could be the economic ruin of this laboratory
- Identity should be known only to a small number of persons
  - These persons must be regularly instructed about there duty to keep this information confidential
- The coordinating body may be required to report poor performance to a particular authority
  - Participants should be notified of this possibility

performance of the laboratories to an authority. In this case the participants have
to agree upon this procedure prior to participation.

Slide 57

Proficiency tests first of all should
help the laboratories improve their
performance. But often they are also
used as a control tool for accreditation
bodies, customers and authorities.
Thus, there may be a tendency among
some participants to try to report a
better performance than is justified.

Collusion and Falsification of Results
- PT schemes often are not only a help for the laboratories to improve their quality but also a control tool for accreditation bodies, customers and authorities
- There may be a tendency among some participants to give a falsely optimistic impression of their capabilities

Slide 58

Collusion between participants must
not be possible. Distributing as many
concentration levels, i.e. different
samples, as possible to confuse potential cheats, can help to prevent this.

Collusion
- Must not be possible

- As many concentration levels as possible

Slide 59

PT samples normally should be
analysed with the same effort as
routine samples. But in reality this
very rarely is the case. An example is
shown in this slide, where a double
measurement was requested in the
proficiency test. During an accreditation
assessment it was found that the
laboratory executed 40 measurements
to be sure of the value. Where possible,
limiting the sample amount to that
necessary for duplicate measurements can help prevent this.

Number of Multiple Measurements

- Example from reality:
  - Routine: single measurement
  - Asked in proficiency test: independent double measurement
  - Executed in proficiency test: 40 (!) measurements
- Therefore: limitation of sample amount, where possible

Slide 60

Usually different levels and different
samples in proficiency tests are evaluated separately. This can lead to
injustice if by chance good laboratories
are grouped together in one concentration level and less good laboratories in
another. This results in variations in
the tolerance limits as shown in this
slide. A calculation procedure for a
variance function was introduced in
the German standard DIN 38402-45
and also published in ISO/TS 20612.

Level-by-level Evaluation for Different Concentrations

- Can lead to injustice

- A procedure for a common evaluation can be found in ISO/TS 20612:2007

Slide 61

What is the effort required for a laboratory to participate in a proficiency
test? First it is the effort to analyse the
sample, which should not exceed the
effort for routine samples but in reality
it is not insignificant. Second it is the
participation fee that the laboratory
has to pay.

Effort for the Laboratory

- Analysis of the samples
  - Should not exceed the effort for routine samples
  - In reality not insignificant
- Participation fee

Slide 62

The participation fee usually is in the range of 300 US-$ and 1000 US-$ depending mainly on the matrix and the parameters that have to be analysed.

Participation Fee

- Normally between 300 US-$ and 1000 US-$ (depending on matrix and parameters)

Slide 63

But there are a lot of benefits for the participants, which are listed in the next three slides.

Benefits - I

- Regular, external and independent check on data quality
- Assistance in demonstrating quality and commitment to quality issues
- Motivation to improve performance
- Support for accreditation/certification to quality standards
- Comparison of performance with that of peers

Slide 64

Benefits - II

- Assistance in the identification of measurement problems
- Feedback and technical advice from organisers (reports, newsletters, open meetings)
- Assistance in the evaluation of methods and instrumentation
- A particularly valuable method of quality control where suitable reference materials are not available

Slide 65

Benefits - III

- Assistance in training staff
- Assistance in the marketing of analytical services
- Savings in time/costs by reducing the need for repeat measurements
- A guard against loss of reputation due to poor performance
- Increased competitiveness

Slide 66

Weighing up costs and benefits, the costs are noticed immediately and the benefits are difficult to quantify, least of all in monetary terms.
But the participation in PT schemes often is an important proof of competence and therefore more than compensates the cost of participation.

Benefits - Costs

- The costs are noticed immediately
- Benefits are difficult to quantify in monetary terms
- The successful participation often is a important proof of competence
- And therefore compensate for the costs of participation

Bibliography

Huber PJ (1981) Robust statistics. Wiley, New York

International laboratory accreditation cooperation (2007) G13 – ILAC Guidelines for the requirements for the competence of providers of proficiency testing schemes, available from http://www.ilac.org

ISO 13528:2005 - Statistical methods for use in proficiency testing by inter-laboratory comparison

ISO/IEC 17043:2010 - Conformity assessment — General requirements for proficiency testing

ISO/TS 20612:2007 - Water quality — Interlaboratory comparisons for proficiency testing of analytical chemistry laboratories

Koch M, Baumeister F (2008) Traceable reference values for routine drinking water proficiency testing: first experiences. Accred. Qual. Assur. 13, 77-82

Koch M, Metzger, JW (2001) Definition of assigned values for proficiency tests in water analysis. Accred. Qual. Assur. 6, 181-185

Lawn RE, Thompson M, Walker RF (1997): Proficiency testing in analytical chemistry. Royal Society of Chemistry for the Laboratory of the Government Chemist (LGC), Teddington

Rienitz O, Schiel D, Güttler B, Koch M, Borchers U (2007) A convenient and economic approach to achieve SI-traceable reference values to be used in drinking-water interlaboratory comparisons. Accred. Qual. Assur. 12, 615-622

Thompson M, Ellison SLR, Wood R (2006) The international harmonized protocol for the proficiency testing of analytical chemistry laboratories. Pure Appl. Chem. 78, 145-196

Index

A

acceptance zone 268
accommodation 35, 151
accreditation 5, 19, 73
accreditation body 20
accuracy 5, 172, 228, 233, 313
action limit 278, 280
affinity diagram 128
algorithm A 315
alternative hypothesis 174
amendments 157
analyte 225
analytical function 187
analytical method 314
ANOVA 309
APLAC 4
apparatus 105
arithmetic mean 164, 315
assessment 325
assigned value 308, 326
audit 6
authority 52
authorization 35
average run length 282

B

base quantities 207
basic calibration 187
benchmarking 128
between-batch condition 276
bias 6, 224, 234, 264
BIPM 2, 3
blank sample 285
blank value chart 285
brainstorming 129

C

calibration 183, 287, 301, 312
calibration certificate 41
calibration curve 13
calibration function 187
cause and effect diagram 129
certificate 305, 321
certification 6, 73
certified reference material 37, 210, 221,
        237, 262, 289, 294, 301, 312
changes of documents 156

CITAC 4
collusion 322
COMAR 300
combined standard uncertainty 16, 253, 264
combined uncertainty 254, 257
commitment 144
communication 53
competence 78
complaints 31, 153
compliance assessment 271
confidence limit 171
confidentiality 320
conformity assessment 76
consensus 317
continual improvement 69, 145
continuous improvement 114, 118
control charts 130, 266, 281
control of measuring device 64
control of production 62
control of records 32
control sample 287
conventional method 314
conventional true value 11
corporate objectives 118
corrective action 31, 69
correlation coefficient 189
cost of conformance 136
cost of non-conformance 136
cost of quality 113, 136
coverage factor 16
critical value 195
CRM *see certified reference material*
CRM production 305
Crosby 132
cumulative distribution 163
customer 84
customer communication 58
customer focus 50
customer property 63
customer satisfaction 116
cusum control chart 281

D

data collection form 128
decision rule 271
defect equipment 152
defective products 139